The Composting Troubleshooter

How to Compost and What to Do If It Goes Wrong

Jane Gilbert

CARBON
CLARITY
PRESS

www.carbon-clarity.com

Published by Carbon Clarity Press

First edition

Copyright © 2015, Jane Gilbert. All rights reserved.

The information provided in this book has been designed to provide helpful information on small scale composting for non-commercial purposes. Although the author and publisher have made every effort to ensure that the information in this book was correct at time of press, the author and publisher do not assume and hereby disclaim any liability to any party for any loss, damage or disruption caused by errors or omissions, whether such errors or omissions result from negligence, accident, or any other cause.

References are provided for informational purposes only and do not constitute endorsement of any websites or other sources. Readers should be aware that the websites listed in this book can and do change.

ISBN: 978-0-9932017-0-7
Version 1.0

CONTENTS

Preface

Over the years, I have spoken to many people about composting. Whilst they nearly always appear to take the benefits of making and using compost for granted, it's the difficulties they've encountered that tend to form the mainstay of our conversations. Comments such as: *"It's just a soggy mess"*, *"After two years I still don't have any compost"*; and *"I can't remove the compost through the little door at the bottom of the bin"*, are not uncommon. So this got me thinking.

In this book I've tried to address these concerns. This book is for anyone who has tried composting and perhaps given up, or is currently composting but is not entirely sure whether it's working as it should. If you've never composted and are thinking of giving it a go, then this book will help propel you up the learning curve, saving you time and hours of frustration.

The Composting Troubleshooter focusses on small-scale backyard (garden) composting, so it's ideal for householders, schools, community gardens, country clubs, allotment holders and the like. Each chapter contains a short introduction at the beginning and a brief summary at the end, so that if you chose to dip in and out to learn about specific topics, it should, hopefully, still make sense.

Throughout the text I've endeavored to smooth over some of the idiosyncrasies between American and British English. Words such as compost 'heap' and 'pile' can be used interchangeably; likewise with 'backyard' and 'garden', and 'fall' and 'autumn'. I've also expressed units in both United States customary units and metric, so you can apply whichever system you're most comfortable with.

Finally, I trust you find this book both informative and useful – if you find the time, please rate it to let me, and others, know what you think.

Happy composting!

Jane Gilbert
Northamptonshire, March 2015

Dedication

This book is dedicated to my Mum, for inspiring me when I was a kid and letting me dabble around in mud; my Dad, who has never composted in his life but thinks it's wonderful nonetheless; and my kids, for providing me with the opportunity to spend countless hours at kids' play barns and ballet classes, allowing me to finally put pen to paper.

1 Introduction

Have you ever been frustrated at how long it takes to produce compost at home? Do you know something's gone wrong with your compost but not sure what? Are you bothered by troublesome flies in your compost bin? Then you're not the only one – these are all common problems experienced by backyard composters the world over. Although there are plenty of books written about how to make compost, few focus on preventing problems and providing advice on what to do when things don't quite go to plan. That's where this book is different.

Composting, in its simplest sense, is a natural process that recycles dead plants and animals, returning their nutrients to the soil so they can be used again and again. You only have to take a walk in a wood and kick around in the fallen leaves to see how they gradually rot down, turning from light to dark brown and then, finally, being transformed into a rich crumbly material that we call compost.

Humans have harnessed this natural process to improve agricultural productivity for millennia, with reports of composting in the Far East reaching as far back as four thousand years. Today, composting can take many guises: from small-scale backyard compost bins, through to large-scale industrial facilities recycling many tens of thousands of tons of material every year. All, however, have one thing in common: they are governed by the laws of nature.

To compost successfully, a blend of both art and science is therefore required. This means having some understanding of the basics, then putting this knowledge into practice. In this book I have explained the key components needed for successful composting. The text is illustrated using simple diagrams, so if you're a visual person, these concepts should be easy to understand.

As with many things in life, putting theory into practice can be challenging. Backyard composting is subject to many variables outside of our control; for example, the weather and seasonal changes in garden debris can markedly speed up or slow down the composting process.

The Composting Troubleshooter aims to bridge this gap between theory and practice. In it you'll learn how to prevent composting problems by blending different materials, getting the structure of the heap right, and adjusting the moisture levels. I've also provided an in-depth description of the different types of materials that can be composted, listing their key properties and things to keep an eye on. You'll also find out about materials that may cause you problems, or taint your compost.

Unlike other composting books, the main focus of this book is to provide you with detailed advice on how to overcome a range of different composting problems. After reading it you should be able to recognize the cause of your problem and make a few simple changes to sort it out. I've also included a Quick Look-Up Troubleshooter Guide to help point you in the right direction, especially if you're in a hurry.

As composting is such a diverse subject, I've also included a chapter on alternative composting methods, such as high-fiber composting, worm composting and Bokashi composting. Hopefully, this should give you a flavor of the range of composting options open to you.

If you follow the advice in this book and manage to make compost successfully, you'll want to know what to do with it. The chapter on putting your compost to work describes some of the many benefits compost can have and the ways in which it can be used. I've outlined how you can apply your compost directly to soil and mix it with other products to use in pots, or make it into a compost tea.

Finally, I've included two short chapters on how to stay safe and healthy when composting, as well as resources you can look up should you wish to learn more.

After reading this book, you'll be able to put your new-found knowledge into practice and go off and be a successful composter. But don't forget: composting is a skill; something that involves an element of trial and error. So, if it things don't quite work out first time, don't be disheartened. Try making a few simple changes, as these can often rectify the problem and get the system back on track. My aim is for both you and your garden to be able to reap the rewards of this wonderful stuff we call compost!

2 Quick Look-Up Troubleshooter

I'd love it if you read this book from cover to cover. It's jam-packed with lots of useful information and advice that can help empower you to make and use compost effectively. However, if you're anything like me, you may not have the luxury of being able to sit down and read it from start to finish. So, if you're just after a quick fix, then here's a troubleshooting summary to help you out.

Nonetheless, please bear in mind that composting is both an art and a science, so even a little bit of knowledge can help you prevent problems occurring in the first place. That's really what this book is all about.

2.1 Problems making compost

HEAP IS TOO COLD OR NOT HEATING UP	
POSSIBLE CAUSES	**SUGGESTED SOLUTIONS**
Heap is too small	Increase the size of the heap to at least 1 yd³ (about 1 m³).
Poor insulation	Cover the sides and top with old carpet or cardboard to conserve heat.
Heap is too dry	Cover the heap to reduce moisture loss. Add water.
Too much structure	Compact the feedstocks.
Too many Browns / insufficient Greens	Blend in some more Greens.
Insufficient air	Turn your compost.
Location of bin is too exposed	Move to a sheltered part of the garden.

COMPOSTING IS TAKING TOO LONG	
POSSIBLE CAUSES	**SUGGESTED SOLUTIONS**
Heap is too dry	Cover the heap to reduce moisture loss. Add water.
Too many Browns / insufficient Greens	Blend in some more Greens.
Insufficient air	Turn the heap. Add more structural material.

HEAP IS TOO DRY	
POSSIBLE CAUSES	**SUGGESTED SOLUTIONS**
Too many Browns / insufficient Greens	Blend in some more Greens.
Too much moisture loss	Cover bin to reduce moisture loss. Add water.
Location is too sunny / windy	Move bin to a more sheltered part of garden.
Surface area to volume ratio is too large	Increase the size of the heap to at least 1 yd^3 (about 1 m^3).

HEAP IS TOO WET	
POSSIBLE CAUSES	**SUGGESTED SOLUTIONS**
Too many Greens	Mix in some Browns.
Poor structure – has become compacted	Open up structure by mixing in Browns e.g. shredded paper.
Poor drainage	Create layer of twigs / branches at bottom of bin to improve drainage.
Inadequate air circulation	Improve structure by adding Browns. Drill holes in sides of bin.

COMPOST SMELLS	
POSSIBLE CAUSES	**SUGGESTED SOLUTIONS**
Too wet	Dry the compost out, add structural materials and improve drainage.
Too many Greens	Mix in some Browns.
Insufficient air	Improve structure by adding Browns. Drill holes in sides of bin.

PLANTS GROWING INTO COMPOST BIN	
POSSIBLE CAUSES	**SUGGESTED SOLUTIONS**
Scavenging nutrients released from the compost	Physically prevent them reaching the bin – move bin or create barrier at bottom e.g. by using chicken wire. Use a tumbler bin.

2.2 Unwanted visitors

VERMIN (RATS & MICE)	
POSSIBLE CAUSES	**SUGGESTED SOLUTIONS**
Attracted by food	Keep cooked food out of the bin, especially meat, fish and dairy products.
Have found a comfortable home – heap is too dry	Increase moisture level inside bin.
Heap forms part of their territory	Move bin to different part of garden. Physically prevent access by placing mesh over bottom of the bin.

FLIES & ANTS	
POSSIBLE CAUSES	**SUGGESTED SOLUTIONS**
Attracted by food	Cover up fruit with barrier e.g. shredded paper.
Have found a comfortable home – heap is too dry	Increase moisture level inside bin.

2.3 Problematic compost

TWIGS & BRANCHES IN FINISHED COMPOST	
POSSIBLE CAUSES	**SUGGESTED SOLUTIONS**
Woody material not fully decomposed	Remove by hand or using a sieve. Leave to compost for a few more months.

UNABLE TO ACCESS COMPOST	
POSSIBLE CAUSES	**SUGGESTED SOLUTIONS**
Finished compost is inside bin covered with undecomposed feedstocks	Remove bin and un-composted feedstocks.

ONLY PART OF THE HEAP HAS COMPOSTED	
POSSIBLE CAUSES	SUGGESTED SOLUTIONS
Heap may have pockets of dry material / poor mix of Greens & Browns	Harvest what compost you can, re-mix remainder and put back in the bin.

WEEDS GROWING IN THE COMPOST	
POSSIBLE CAUSES	SUGGESTED SOLUTIONS
Viable weed roots, rhizomes and seeds put in bin	Remove weeds by hand (if large). Dig compost into soil. Turn the compost to break up roots. Dry the compost out to kill off weeds, then re-compost.

'DUSTY' COMPOST	
POSSIBLE CAUSES	SUGGESTED SOLUTIONS
Growth of composting microbes on surface ('fire fang')	Water compost before turning or moving to reduce dust.

COMPOST HAS KILLED PLANTS	
POSSIBLE CAUSES	SUGGESTED SOLUTIONS
Compost is not mature enough	Leave it to compost for a few more months.
Compost is unstable causing nitrogen 'lock up'	Leave it to compost for longer. Add some more Greens.
Herbicide residues remain	Leave compost for at least another year.

COMPOST IS FULL OF CREEPY CRAWLIES	
POSSIBLE CAUSES	SUGGESTED SOLUTIONS
Compost provides them with a home and food	Remove compost from bin and leave in open air for a few hours to disperse them.

3 Composting Basics

All composting systems, irrespective of their size, depend upon a range of different organisms to break down waste materials and convert them into compost. Some are involved in the early stages of decomposition, feeding off the vegetation we pile into our compost heaps, whilst others come into play during the latter stages, eating up the by-products of other organisms who have already eaten their fill. The first part of this chapter provides a brief outline of what these organisms are and the important work they do.

The rest of this chapter looks at the main variables that affect how efficiently and effectively these compost organisms work. There are sections on the types of food they like to eat, how much air and water they need, as well as the temperatures they prefer. It's helpful to learn about these basic variables, as when one or more get out of kilter, problems may well arise.

3.1 Compost biology

Microbes

All composting, irrespective of whether it is carried out on a large industrial scale or in a small back yard, relies upon a diverse range of microorganisms ('microbes' for short). These are tiny organisms that are generally too small to see with the naked eye. They are found everywhere on earth, from frozen polar regions to hot, dry deserts, and even in the guts of us humans, where they are responsible for recycling nutrients and organic matter.

The microbes in a compost heap are a diverse range of different bacteria and fungi that eat up waste materials, using some of it as a food to fuel their growth and reproduction, whilst at the same time creating a humus-rich by-product called compost. Different types of microbes grow in a compost heap during different stages of the process: some are better adapted to eating fresh materials at the outset, whilst others prefer partially decomposed materials left over after the initial microbes have satiated their appetites.

Temperature also influences the types of microbes in a compost pile. Like us, active microbes release heat energy as they feed, which can increase the temperature of the composting mass. Microbes that prefer warm, but not hot, temperatures (somewhere between 68 and 113 °F or 20 and 45 °C) are termed 'mesophiles' (or middle temperature lovers). They colonize compost heaps at the beginning and end of the composting process and include both bacteria and fungi.

Microbes that prefer high temperatures (above 104 °F or 40 °C) are called 'thermophiles' (or heat loving), so these are the dominant players in commercial and large scale composting systems, which can get quite hot. During thermophilic composting it is the bacteria that dominate.

As most garden compost heaps do not heat up appreciably above ambient temperatures, this means that the mesophiles play an important role in backyard composting. Fungi, in particular, are important composting microbes, as they are adept at degrading woody materials.

Composting creepy crawlies

In addition to the bacteria and fungi, there is also a complex web of larger organisms that feed off these microbes and their products. As they are unable to survive at high temperatures, these creepy crawlies will only colonize compost heaps if they do not heat up too much, or after they have cooled down. They are important during the latter stages of composting, helping to mix and break the organic matter into smaller pieces. Additionally, their feces (poop) are also a good source of plant available nutrients, rich in beneficial microbes.

In general, compost creepy crawlies can be divided into two main types:

- **The primary consumers** – these feed off the composting microbes and dead pieces of vegetation. They include the earthworms, ants, millipedes, flies, slugs and snails; and
- **The secondary consumers** – these are predators that feed off the primary consumers, and include beetles, centipedes and spiders.

10

The interrelationship between microbes, primary and secondary consumers is shown in the diagram below. In my view, it is a complex and fascinating subject, which deserves further study.

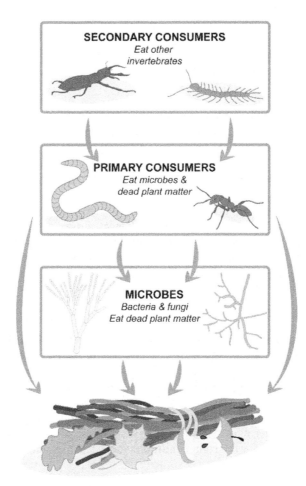

Simplified compost food chain

Now that that we've introduced some of the main composting organisms, let's look at the foods they need and how they affect the composting process.

3.2 Blending materials

As we've just learned, composting relies upon a myriad of microscopic organisms to break down waste materials, which they use as a source of food. And, just like us humans who need a balanced diet to help keep us healthy, these microbes do too. In the same way that too much cake, or too few vegetables can make us ill, if the mix of materials in a compost heap is not balanced, then the whole system can go off kilter. Getting the mix right will help you run an efficient composting heap – that means making good quality compost in a timely manner.

So how should you go about it? Basically, waste materials can be divided into broadly two types: the 'Browns' and the 'Greens'. In food terms, the Browns can be likened to muesli (rich in fiber and providing a slow-release energy source) whilst the Greens can be likened to cake (rich in protein and full of fast-burn sugars).

A balanced mix of both Browns and Greens is what is needed. However, that is where the challenge lies, as most gardeners tend to have a surplus of one or another at different times of the year. Before I discuss this in more detail, let's take a quick look at the main features of both types of material.

The Browns

The leaves that fall off trees, and the wood that makes up their trunks and branches, are the slow-burn foodstuffs on which a composting heap relies upon. These 'Browns', are gathered up in abundance by us gardeners, especially during the fall and winter months.

The Browns are physically very strong and, when dry, are also resistant to decay. This is why we use wood in our homes and buildings, as under the right conditions it can last for centuries.

The Browns are rich in carbon, but low in nitrogen. On their own, and given enough moisture, they will rot down slowly, forming dark-brown organic matter-rich compost. Depending upon the climate, this could take many months, if not years. They are a slow and steady source of energy for a compost heap, forming the basis of the all-important humus, which does so much to improve the structure of our soils. The main features of the Browns are shown in the box below.

MEET THE BROWNS	
CHARACTERISTICS	SOURCES
▓ Woody ▓ Tend to be dry (low in moisture) ▓ High in carbon ▓ Low in nitrogen ▓ Branchy (take up a lot of space) ▓ Slow to rot down	▓ Tree trunks and branches ▓ Fallen leaves ▓ Straw ▓ Shredded paper and card

Composting Browns on their own will therefore take a long time. In itself there's nothing wrong with this, but let's face it, you will probably want to put your compost to good use as soon as possible. It's a valuable resource, so why wait? There's also the issue of space. Few of us have enough space in our gardens to stockpile all these pruned branches and twigs and wait until they have miraculously decomposed.

So what's the answer? Well, in practical terms you'll need to speed up your composting heap by adding some more easily digestible foods. This is where the 'Greens' come in.

The Greens

The Greens are the 'high-energy' foods that act as a booster for your compost heap. They are the grass clippings, flower heads, stems and discarded vegetables that are especially plentiful during the late spring and summer months. The Greens tend to be soft and sappy, are high in moisture, and (here's the important element), they are also high in nitrogen. This nitrogen is needed by the composting microbes, which use it as a nutrient during the composting process. The main features of the Greens are shown in the box below.

MEET THE GREENS	
CHARACTERISTICS	**SOURCES**
▓ Soft and sappy ▓ Tend to be moist ▓ High in nitrogen ▓ Compact easily ▓ Rot down quickly	▓ Grass clippings ▓ Flower heads and stems ▓ Bedding plants ▓ Discarded vegetables and fruit

Composting Greens on their own is generally not a good idea. Their sappy nature means they can compact easily, forming a dense soggy mass. This, in turn, effectively prevents air from circulating inside the materials, which may lead to the formation of nasty smells. (More on this in a bit.) In addition, the high nitrogen content can also exacerbate this problem.

Balancing Browns and Greens

To run efficiently, a compost heap therefore needs a decent mixture of slow and fast burn foods – that is, a balance of Browns and Greens, with the Browns providing the bulk of the carbon, and the Greens the nitrogen.

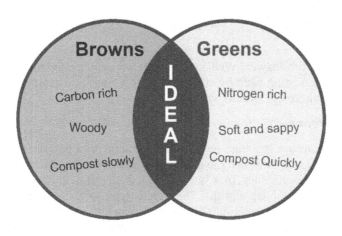

A balance of Browns and Greens works best

Commercial composters aim to blend their materials to achieve an optimal balance of carbon and nitrogen. The relative proportion is termed the carbon-to-nitrogen ratio, or C:N for short. Ideally, a compost heap needs to be supplied with materials where the C:N ratio is somewhere between 25:1 and 40:1. (This means that for every pound or gram of nitrogen there should be between 25 and 40 pounds or grams of carbon.) When the ratios are correct, there should be sufficient nitrogen present for the composting microbes to break down the woody, carbon-rich material and form compost efficiently. If there is proportionally too much carbon compared to nitrogen, then the rate at which composting occurs will be slow. If there is too much nitrogen relative to carbon, then the excess nitrogen will be released into the air, either as ammonia or other odorous substances.

So how do you know how much carbon and nitrogen there are in your materials? Well, there are a number of scientific publications, websites and even smart phone apps that list the C:N ratios of different materials to assist budding composters select the correct blends. Spreadsheets can then be used to work out specific ratios. If that is of interest, then further information is listed in Chapter 9.

But "hold on a sec", I hear you saying, "this is way, way too complicated"! And yes it is; backyard composting doesn't need to, and shouldn't, extend to this level of sophistication. Let's leave spreadsheets and calculations for those with a real passion, scientific researchers or commercial composters.

The important thing to remember is that materials destined for your compost heap can be broadly classified into two main types: the Browns and the Greens, and a balance of both is needed for effective composting to take place. There's no need to measure them and work out ratios mathematically. Instead, you'll get a feel for how well these materials compost in your own garden and learn what works and what doesn't work. In Chapter 4 I've listed the main materials that may be generated in a typical yard and composted, and classified them as either a Brown or a Green, to help get you started.

3.3   Getting the structure right

All compost heaps need air. Air, or rather the oxygen in it, is used by the composting microbes in the same way that we humans need it. The main difference is that we have lungs to actively extract the oxygen from the air, whereas a composting heap relies on air moving in and out of the pile through natural convection currents. This means that there need to be sufficient air channels in the composting mass to allow this to happen.

Density is important

It follows that if your composting materials are too small and too densely packed, then very little, if any, air can circulate through them. On the other hand, materials that are too branchy and create large air pockets allow lots of air to circulate freely. However, these heaps will have difficulty retaining heat and will dry out far faster than denser heaps. In practice a balance between these two extremes needs to be struck.

Structure is too dense	Structure is just right	Structure is too branchy
Small air pockets	Air can circulate	Too much air circulates
Poor air circulation	Moisture & heat are	Materials dry out too
Can turn smelly	retained	quickly
		Unable to retain heat

Density is the key

Density is a measure of how closely packed the composting materials are, and will vary according to how old the compost heap is. So what is the optimal density? Text books tell us that a density of round about 1,000 lb / yd^3 (or about 600 kg / m^3) is optimal, which is fine in principle, but of little relevance to you as a gardener composting at home.

In practice, the easiest way to ensure that your compost heap is of a reasonable density is to make sure that there is a decent mix of Browns and Greens. As the Browns generally consist of woody materials, they provide a framework of air pockets, allowing gases to circulate. The Greens will provide the bulk of the moisture and nutrients, so will start to decompose fairly rapidly, allowing composting to get started.

AN UNBALANCED MIX	
TOO MANY BROWNS	**TOO MANY GREENS**
▦ Branchy structure ▦ Not dense enough ▦ Dries out quickly ▦ Loses heat easily	▦ Compacts easily ▦ Compost becomes too dense ▦ Air cannot circulate easily ▦ Oxygen becomes limited ▦ Stale gases build up ▦ Can become smelly

Density can also be managed by chopping up materials; this helps to reduce their size and make them more manageable. It can be done either manually or with the help of a garden chipper or shredder. Chippers or shredders can quickly reduce the size of larger woody materials, thereby saving space and helping to kick-start the composting process. However, they can be expensive to buy, and, unless sized appropriately, they may struggle with larger branches.

Chipping also has the added advantage in that it increases the surface area of the materials, meaning that more of the material is exposed to the composting microbes. The smaller the particles, the greater the surface area. However, remember that a balance needs to be struck between maximizing surface area for degradation to begin, whilst maintaining sufficient structure to allow air to circulate.

PARTICLE SIZE RULE OF THUMB
Particle sizes of between 1 – 2 inches (25 – 50 mm) are ideal.

So, does this mean you have to measure the density and particle sizes of your composting materials? No, of course not. What it does mean though, is that if you have a lot of branchy Browns, chop them up if possible. If it's not possible, don't worry. It just means that things may take a little longer. If you have a lot of dense, compacted materials such as grass clippings, then blend them with a different material to provide some structure and to prevent them smelling.

How do you know if you've got it right? The best way is to set your materials off composting, then check them after a month or so. If they appear damp and look like they've started to rot down, chances are you've got it right. If they look just like they did on day one, then chances are they may need compacting a bit and additional water adding (see below).

Compacting your compost heap can be done quite simply by breaking up the materials with a garden spade or shovel (but take care there aren't any hibernating or nesting animals in your heap), or by standing on the top of the heap and jumping up and down on it (but please be very careful not to hurt yourself!).

Introducing air

However ideal the structure of your composting materials at the outset, as they begin to decompose and reduce in size, internal air pockets become filled in, the structure starts to slump, and as a consequence, movement of air will become impeded. To overcome this, the composting materials can be picked up and mixed through a process called 'turning'. Turning helps to:

- Introduce fresh air into the composting mass;
- Release stale gases;
- Mix the materials up, blending the Greens and Browns even further, and helping to transfer drier materials on the outside of the heap into the damper center; and
- Re-create new air pockets and improve the structure of the partially composted material, thereby improving gas circulation.

In commercial composting systems, turning can involve large, complex machines costing six figure sums of money. Fortunately, gardeners can achieve this at a fraction of the cost using a garden fork and a little bit of muscle.

So what is involved? In its simplest sense, it involves breaking up the compost heap, picking up partially composted material and then dropping it back down, trying to mix it as much as possible. If you're using a plastic compost bin, it's best to lift this up and remove it before turning. If your bin is of a homemade variety, then turning is easier if you're able to remove the front panel to gain access to the compost.

Turning can also provide a good opportunity to move it to a different bin or container. It is also an ideal time to add water to the composting materials if they appear to be too dry, or to let them dry off in the sun if they are particularly wet.

Some gardeners turn their compost at pre-determined intervals, whilst others never bother, simply letting nature take its course at its own rate.

Turning garden compost is not essential, although it will help speed up the process. It's really a matter of priorities, and whether you need the compost sooner or later.

Unturned heap Small pieces compact at the bottom Layers form Poor circulation of gases	**Turned heap** Materials are mixed up Air channels re-form Gases circulate

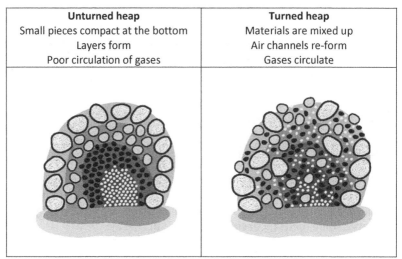

Turning helps to mix the materials

Too little air

If a compost heap doesn't have enough air circulating through it, then two things can happen:

- Composting will slow down; and
- It may start to smell.

Neither are particularly helpful, and should be avoided wherever possible. The following box explains the problem of a lack of oxygen in more detail.

PROBLEMS CAUSED BY TOO LITTLE OXYGEN	
SMELLS	Low or oxygen-free atmospheres support the growth and proliferation of so-called 'anaerobic' bacteria. These are specialized bacteria typically found in bogs, sewers and soils. Unfortunately for us, these anaerobes can release both methane and odorous chemicals, which is the reason why our sewers smell so bad to us. Odors can range from a rotten egg smell (which is typically formed in very wet, compacted compost) to silage-type smells, caused by densely packed grass clippings.
METHANE	Methane is a gas that is commonly called 'natural gas', and is piped into our homes as a fuel for heating and cooking. However, when released directly into the atmosphere it is a potent greenhouse gas, with a global warming potential 72 times greater than carbon dioxide (over a 20-year time span). This means that it contributes towards unwanted climate change – something that goes against the grain for the majority of composters.

3.4 Adjusting moisture

Why moisture is needed

In a compost heap, water acts as a medium for the composting microbes to feed and breathe in. All microbes need some moisture, as without it they are unable to eat or grow. This is the reason why we preserve some foods by drying them – it halts decay.

As microbes don't have mouths or stomachs like animals, they need to digest their foods outside of their bodies before they can absorb the goodness needed to keep them alive. To do this they need to be coated in a thin layer of water into which they secrete digestive enzymes. This means that if there is too little moisture in your compost heap, then the microbes will be unable to do their job. They simply won't be able to feed.

Water is also needed to dissolve oxygen from the air, so that they can 'breathe'. Too little moisture and growth will be impeded as the microbes will have insufficient oxygen.

Conversely, too much water can also be problematic. As we've discussed in the previous section, air pockets within a compost heap are needed to allow gases to move around freely within the composting mass. If there is too much water in the compost heap, then these pockets will fill with water, and prevent air from circulating. Wet, soggy piles can soon start to smell.

Materials are too wet	Materials are just right	Materials are too dry
Air pockets fill with water	Materials are moist, but not too wet	Microbes cannot grow and feed
Poor air circulation	Air can circulate	Composting stalls
Can turn smelly		

Damp composting materials are best

How moist should it be?

Gauging how damp a compost heap should be is an art in itself. It is governed by factors that are largely outside of any gardener's direct control: the weather and the waste materials themselves. The type and size of bin will also affect the rate at which moisture is lost from the heap. All you can realistically do is assess what you have, then decide whether it is too dry, too wet, or just about right.

So, how do you know what is right?

If it is:

- **Just right** – it will look and feel damp. You'll be able to see that the materials have started to change color, first turning into a light, then, gradually, into a dark brown.
- **Too wet** – it will look wet and feel soggy. The chances are that it will also smell.
- **Too dry** – it will look and feel dry. It will also be obvious after a few weeks that the materials aren't composting, looking very similar to the way they did on Day 1.

When materials are first placed in a compost bin, most of the moisture will be locked up inside the plants in the leaves and stems, so it can be hard to judge whether the levels are optimal. Rather than try and analyze it, it's best to simply assess whether you have a reasonable mix of Browns (which tend to be drier) and Greens (which tend to be wetter).

Once composting has started and the materials have begun to break down, then, if you wish, you can use the compost 'squeeze test'. This is a simple way of assessing whether there is sufficient free moisture coating the composting materials to enable the microbes to do their job efficiently. It is a tried and tested method used by professionals and amateurs the world over. The basics are detailed in the box below.

MOISTURE SQUEEZE TEST

Take a handful of partially decomposed materials (wear gloves) and squeeze it tightly for ten seconds. If the material is:

- **Just right** – only a few drops of water will appear and the materials will remain in a ball with the consistency of a sponge once you've opened your hand.
- **Too wet** – water will seep out of the materials and they will remain clumped together once you've opened your hand.
- **Too dry** – you won't be able to see any drops of water and the composting materials will fall apart when you open your hand.

Further advice on managing moisture and what to do if you have too much or too little is given in Chapter 4.

3.5 Temperature

In general, it is difficult to achieve high-temperature ('thermophilic') composting in a garden compost heap for two main reasons:

Firstly, a lot of fresh material needs to be composted all at once. In general, volumes in excess of 1 yd³ (about 1 m³) will be needed, as large numbers of active microbes are required to generate sufficient heat energy to up the temperature appreciably. However, this can be a tall order, as most gardeners (myself included) rarely have sufficiently large volumes of waste materials at any one time. Instead, most people add materials to a heap in a piece meal fashion, dropping in a few pruned stems and clippings as and when they arise.

Secondly, the size of the compost pile can also influence how warm the composting materials get. It is generally the case that composting is faster in a large bin compared with a small bin. Why? Because large compost heaps have a larger thermal mass than smaller ones, which means that the heat energy generated during composting is generally lost at a slower rate in a larger bin than in a smaller one. Once warm, larger bins tend to hold their heat longer than smaller ones.

In garden compost bins, peaks in temperature tend to be relatively short lived, compared to large commercial heaps where temperatures above 158 °F (70 °C) can be reached for many weeks at a time. Low temperature composting is therefore best suited to gardens where small volumes of materials are generated and added to the heap on an ad hoc basis.

If you're still interested in achieving high temperatures in your compost heap, then specifically designed 'hot' compost bins can be purchased. These offer improved insulation over standard plastic bins, and often also come with a 'booster' mix to help kick-start the composting process. Long reach temperature probes can also be bought, so that you can measure and monitor temperature.

Some of the pros and cons of high and low temperature composting are shown in the box below.

	HIGH-TEMPERATURE COMPOSTING	LOW-TEMPERATURE COMPOSTING
PROS	▪ Helps kill off weeds and disease-causing microbes ▪ Produces compost relatively rapidly (within a few months) ▪ Will deter vermin	▪ Simple and easy ▪ Does not require active management ▪ Useful where there are more Browns than Greens ▪ Useful where small quantities of materials at composted on an ad hoc basis
CONS	▪ Requires active management ▪ May need to add moisture ▪ May need to insulate the bin ▪ Requires a large volume of material to be composted all at the same time	▪ May not kill off weeds and diseases ▪ May take a long while to produce compost (few years)

Pros and cons of high and low temperature composting

3.6 Types of compost bin

Compost bins are widely available; they can be purchased at garden centers, home improvement (DIY) stores, over the internet, or directly from your municipality, which may offer subsidized promotional models. They also come in a range of different styles, sizes and colors; most are made out of plastic, some out of wood, and some of the more costly versions are even made out of metal. Alternatively, rather than spending money, you may wish to make your own using readily available materials: old wooden pallets, or posts and chicken wire, are cheap and cheerful options. Examples are shown below.

Budget Simple, homemade bin	Mid-range Purchased static bin	Expensive Purchased tumbler bin
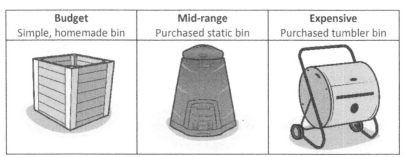		

Different types of compost bin

Your choice of compost bin will be dictated not only by style and personal preference, but also a number of practical considerations, such as:

- **Size** – will the bin be sufficient for your needs, bearing in mind the size of your garden and seasonal fluctuations in garden waste that arise throughout the year?
- **Price** – how much are you willing, or able, to spend?
- **Plastic or wood** – do you have a preference?
- **Homemade or bought** – do you have the time, skill and resources to make one?
- **Availability** – do you need to use one immediately, can you wait for a delivery, or do you have the time to make one?
- **Climate** – do you live in a particularly dry or wet area, where either keeping moisture in or out will be important? Do you live somewhere particularly cold, where retaining heat in the compost heap will be necessary to prevent it freezing?
- **Speed** – how quickly do you need to make compost?
- **Vermin** – do you live in an area where there are urban foxes and rodents that need to be kept out?
- **Position** – will the bin be tucked away at the far corner of the garden out of sight, or will you need to site it close to the house?
- **Looks** – is the style and visual appearance of the bin important to you, given the style and look of your garden?

3.7 Summary

In this chapter we've looked at the different types of organisms that are the powerhouse of a compost heap, and the three main components that they need to work efficiently and effectively. As a gardener, you can help these composting organisms by:

- Blending materials so there is a balance of Browns and Greens;
- Ensuring there is an adequate air supply to enable the composting microbes to 'breathe' and to prevent the heap smelling; and
- Ensuring that there is a thin film of moisture covering the feedstocks, so that the composting microbes have enough water to digest their food.

These vital components are shown below:

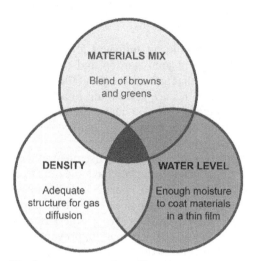

Vital components for effective composting

Getting all three variables within the desired ranges all of the time is a tall order. However, with a bit of cajoling and skill, it is possible to keep most of these them within acceptable levels most of the time.

We also briefly discussed temperature and how hot heaps generally decompose faster than cold ones. As most garden compost heaps are too small to retain much heat, cool heaps should not be any cause for concern.

Finally, although there is a bewildering array of compost bins available to purchase, before investing any money, it's worth considering your needs and preferences.

4 Prevention is Better than Cure

In the previous chapter we've looked at the three vital components needed for efficient and effective composting. However, as most gardeners are only able to compost waste materials as and when they arise, this means that at certain times of the year there will most probably be a surplus of Greens (especially during the summer months) and then Browns (during fall and winter). By understanding the nature of these materials and being aware of their properties, this should help you prevent problems occurring in the first place.

In this chapter I've described some of the common waste materials that gardeners may encounter over the course of a year, along with any properties that may lend themselves to troublesome composting. In addition, knowing what to keep out of a compost heap is just as important as knowing what to put in, so I've also listed some materials that are best avoided and the reasons why.

This chapter also includes a short section on composting recipes, as well as suggestions about how to deal with surplus materials when it is not possible to obtain a balanced mix of Greens and Browns. As they say "prevention is always better than cure"!

4.1 Types of compostable materials

The basic properties of a range of different compostable materials have been listed here in alphabetical order. I've endeavored to describe some of their typical characteristics alongside any features that may be troublesome in a compost heap. The aim of this part of the book is to provide a reference for you to dip into and out of as and when required.

Bedding plants

MATERIAL TYPE (BROWN OR GREEN)	STRUCTURE (AIR CIRCULATION)	MOISTURE (CONTENT)
Green	Good	High – Low

If you like to decorate your garden or window boxes with annuals, then at some stage you'll need to remove them after they have flowered to make way for next season's display. The moisture content will be largely dependent upon whether they are still alive and have just finished flowering, or whether they have died off and dried out *in situ*. Either way, they should be readily compostable, although it's best to remove any excess soil adhering to the roots before placing them in the compost heap.

Bracken

MATERIAL TYPE (BROWN OR GREEN)	STRUCTURE (AIR CIRCULATION)	MOISTURE (CONTENT)
Brown / Green	Good	Medium – Low

Bracken is a fern that isn't generally grown by gardeners, although it can migrate into gardens from neighboring countryside or woodlands, where it may start to outcompete more favorable garden plants. In addition to being invasive, the fronds and spores are thought to be carcinogenic, especially if eaten. Don't worry though; it isn't thought to present any significant danger to gardeners handling it on an ad hoc basis.

Bracken can be classed as either a Green or a Brown, depending upon whether it is green and actively growing, or has started to die back at the end of the season and turned brown. If possible, it's best not to remove it during hot, dry weather, as this is when the spores tend to be released.

Bracken rots down to create a useful compost which is slightly acidic (ericaceous) and good for lime-hating plants. It is also a good source of potassium.

Cardboard and paper

MATERIAL TYPE (BROWN OR GREEN)	STRUCTURE (AIR CIRCULATION)	MOISTURE (CONTENT)
Brown	Good (if shredded or scrunched up)	Low

Paper and cardboard are good substitutes for garden-grown Browns, and are also a main ingredient in high-fiber composting recipes (see Section 6.4). If you plan to use paper or card, it is best to shred or scrunch it up first, otherwise it may well stick together and form a big lump once it has become damp. (This then takes ages to compost, which defeats the point of using it in the first place).

Additionally, take care not to use too much colored or glossy paper, as the inks may potentially contaminate the compost. Cardboard can also contain a range of non-biodegradable glues, tapes, staples, laminate coatings and plastic document envelopes; all of these are best avoided.

Coffee grounds and tea leaves

MATERIAL TYPE (BROWN OR GREEN)	STRUCTURE (AIR CIRCULATION)	MOISTURE (CONTENT)
Green	Poor	Low - High

Both tea leaves and coffee grounds are a good source of nitrogen, and can act as a booster for a compost heap. Some percolator coffee pots use paper filters, and these can be composted along with the grounds. It is said that large quantities of coffee grounds can turn the compost acidic, although some research reports have cast doubt on this. If you source any significant quantity of coffee grounds, such as from a local caterer, then it's probably best to make sure they form part of a well-mixed compost blend, including plenty of Browns.

Whilst tea leaves are a valuable addition to any compost heap, the bags that hold the leaves can be problematic. Some bags are made entirely out of plastic, which don't compost down. These are best avoided, or, better still, split open to remove the tea leaves (if you have the time and patience), then disposed of. Others, on the other hand, may be made out of paper that has been lined with a thermoplastic. This plastic can take a long time to degrade, and may be visible in a compost heap after the rest of the material has decomposed. These little 'skeleton' bags can then be easily removed from the finished compost and disposed of. Some tagged tea bags may also have the string attached with a staple. I generally don't worry about this as the staples are so small and tend to rust fairly readily, providing a useful source of iron.

Food waste

MATERIAL TYPE (BROWN OR GREEN)	STRUCTURE (AIR CIRCULATION)	MOISTURE (CONTENT)
Green	Poor - Medium	High

There are generally two main types of food waste produced by households: unavoidable and avoidable. Unavoidable food waste is the stuff left over during food preparation and includes, for example, vegetable peelings, cut stalks and egg shells. Avoidable food waste is food that could otherwise have been eaten, and includes plate scrapings, out-of-date foods, bread crusts etc.

Conventional wisdom says keep cooked foodstuffs out of a compost heap to prevent them attracting vermin. However, in my experience, there is very little difference between a cooked potato and an uncooked potato peeling as far as a hungry rodent is concerned. The main difference is whether the food contains **meat, fish or dairy products. These should definitely be avoided**.

In Europe, food that comes out of a kitchen falls under the scope of the **EU Animal By-Products Regulations**. Different European Member States have interpreted these regulations in different ways, with the UK, in particular, taking a strict view. If you keep ruminant animals, pigs or poultry on the premises where you plan to compost food wastes, then it is best to seek advice from your relevant authority.

Egg shells are about 95% calcium carbonate (the same substance in chalk and limestone), and therefore do not compost *per se*. However, they are useful amendments in any compost heap, helping to add calcium to the soil. They may, however, remain visible in the compost and may need to be crushed to reduce their particle size. Similarly, although technically biodegradable, **avocado stones** and **nut shells**

may also take a long while to rot down. If you find them in your finished compost, then simply remove and recirculate in an active compost heap.

Fats, oils and greases

MATERIAL TYPE (BROWN OR GREEN)	STRUCTURE (AIR CIRCULATION)	MOISTURE (CONTENT)
Brown	Poor	Low

Fats, oils and greases (FOGs) are rich in energy, and will be rapidly degraded by composting microbes. It is important only to use FOGs if they come from a kitchen / food source, although please bear in mind the advice regarding meat and fish wastes and the potential to attract vermin.

Never use waste oils from a vehicle engine or other mechanical device. Why? Firstly, these oils are likely to be synthetic mixes containing additives to improve engine performance. As such, they may not be easy to compost. Secondly, and this is especially true of used car engine oil, they are likely to be contaminated with chemicals that can harm humans, plants and the soil. Keep these oils out at all costs and dispose of through an approved route, such as a municipal collection scheme.

Fruit

MATERIAL TYPE (BROWN OR GREEN)	STRUCTURE (AIR CIRCULATION)	MOISTURE (CONTENT)
Green	Poor	High

Left over fruit may be mixed with other kitchen waste during food preparation, and is a good source of nitrogen and moisture. Additionally, during late summer and fall there may be significant quantities of damaged windfall fruit, as well as cores, pips, stones and peelings if you are making jams or chutneys. It's best to try to cover this waste fruit with some Browns (shredded paper is particularly good), especially if the weather is warm, to avoid them attracting flies and wasps.

Grass clippings

MATERIAL TYPE (BROWN OR GREEN)	STRUCTURE (AIR CIRCULATION)	MOISTURE (CONTENT)
Green	Poor	High

Most gardens have lawns, which means that during late spring and summer, significant quantities of grass clippings can be generated. Grass is a very useful compost amendment; however, it also needs to be managed. Too much grass piled up on its own can heat up very quickly. Two things may then happen: it may dry out and form a compact mat which may then be hard to break up, or it may turn into a smelly, slimy mess.

It's best to try to prevent this happening in the first place by mixing it with Browns, or scattering it loosely in the compost bin to prevent compaction. It's also best not to store grass clippings (e.g. in plastic bags) before putting them on the heap, as they will start to smell (see my 'Salutary Lesson' in Section 5.3 for what can go wrong).

If you treat your lawn with proprietary herbicides, please read the instructions first. Some herbicides such as 'clopyralid' can take many months, if not years, to breakdown. Consequently, it may remain active in the finished compost, where it can then harm plants grown in your compost. This is discussed further in Section 4.2.

Hay and straw

MATERIAL TYPE (BROWN OR GREEN)	STRUCTURE (AIR CIRCULATION)	MOISTURE (CONTENT)
Brown	Good	Low

Hay and straw will add structure, but will generally decay slowly. It is unlikely that most gardeners would wish to compost hay or straw unless it has been used as bedding material for pets, poultry or livestock. In this case, it will be ready mixed with manure (which is classed as a Green and is a wonderful source of nitrogen) and may not need blending with other materials.

If you are buying-in hay, it's worth checking its source and enquiring whether it has been treated with an herbicide that may persist in the compost (again, see Section 4.2).

Hedge trimmings

MATERIAL TYPE (BROWN OR GREEN)	STRUCTURE (AIR CIRCULATION)	MOISTURE (CONTENT)
Brown / Green	Good	Medium

Hedge trimmings will provide useful structure in any compost heap. If you have pruned an overgrown hedge, then the branches may need chopping up further before composting.

Herbaceous border plants

MATERIAL TYPE (BROWN OR GREEN)	STRUCTURE (AIR CIRCULATION)	MOISTURE (CONTENT)
Brown / Green	Good	Low - Medium

If you have an herbaceous border, then at some stage it will need thinning out, pruning or even completely renovating. Either way, you'll most likely generate large quantities of plant material, some of which may well be soft and sappy, whilst some (such as lavender) may be quite woody.

On the whole, these will tend to be easy to compost, needing no special attention, other than shaking off surplus soil adhering to roots if the whole plant has been removed. Larger woody plants may need chopping up before placing in the compost heap.

Leaves – deciduous autumnal fall

MATERIAL TYPE (BROWN OR GREEN)	STRUCTURE (AIR CIRCULATION)	MOISTURE (CONTENT)
Brown	Medium	Low - Medium

Fall is a special time and nature's way of recycling nutrients and ridding deciduous trees of leaves that may otherwise damage it during periods of snow fall or strong winds. During the spring and summer months, leaves are actively photosynthesizing: converting carbon dioxide in the atmosphere into sugars for the tree to feed on. Green leaves are therefore quite high in nutrients and water, and should be regarded as a 'Green'.

The nutrients in the leaves are too valuable for the tree to lose, so before dropping them during fall it reabsorbs them back into the trunk and roots for use in the upcoming year. Autumnal leaves are therefore low in nutrients and moisture, but high in carbon, and can be regarded as 'Browns'. They can be gathered and left to decompose on their own, forming leaf mold (see Section 6.1), or mixed with Greens. Alternatively, they form a useful 'bank' of Browns to use during the summer months when larger quantities of Greens (such as grass clippings are generated).

If you are gathering fallen leaves from areas where dogs are walked, take care, as they may be contaminated with feces and litter. In addition, street sweepings may also be contaminated with litter and residues from road traffic, such as plastics and heavy metals. The key here is to know where the leaves come from and to consider where the compost will be used.

Leaves - evergreens

MATERIAL TYPE (BROWN OR GREEN)	STRUCTURE (AIR CIRCULATION)	MOISTURE (CONTENT)
Brown	Poor	Low

The leaves of evergreen trees differ from those of their deciduous cousins in a few important ways: firstly, apart from a few exceptions, they do not fall off the tree during autumn, but are shed continuously throughout the year. Conifers typically have 'needle' shaped leaves that help the tree shed snow during winter. They also have a waxy coat that helps reduce moisture loss and withstand cold temperatures. This means that evergreen leaves generally do not provide a great deal of structure in a compost heap, but the nature of the leaves means that they may also take a good while to break down.

Manure - horse

MATERIAL TYPE (BROWN OR GREEN)	STRUCTURE (AIR CIRCULATION)	MOISTURE (CONTENT)
Green	Poor	Medium - High

Well-rotted horse manure is a useful addition to any flourishing rose bed. The focus here is on 'rotted', which means that fresh manure needs to be allowed to partially decompose before it is used. This can be done in either a dedicated compost heap, or by mixing the manure with other composting materials, to produce nutrient-rich compost. Again, take care if the horse has grazed on a pasture treated with herbicide, as this may be transferred through the animal's gut and into the manure (see Section 4.2).

Manure - poultry

MATERIAL TYPE (BROWN OR GREEN)	STRUCTURE (AIR CIRCULATION)	MOISTURE (CONTENT)
Green	Poor	Variable

The recent interest in keeping backyard chickens has resulted in the inevitable: poultry manure, which is rich in nitrogen and a useful fertilizer. However, as fresh manure can be smelly and may contain germs, it is wise to compost it first. The manure is, in itself, poor in structure and may be moist or dry, depending upon how the birds have been housed and fed. Fortunately, most manures are mixed with bedding materials (such as sawdust, hay or straw), which adds useful carbon (Browns) and helps absorb excess moisture. Manure on its own will need to be mixed with a good source of Browns in a suggested ratio of one part manure to at least two parts Brown.

When handling any manures, please follow good hygienic practices, as described in Chapter 8.

Moss and lawn scrapings

MATERIAL TYPE (BROWN OR GREEN)	STRUCTURE (AIR CIRCULATION)	MOISTURE (CONTENT)
Brown - Green	Poor	Low

Many gardening books advise us to 'scarify' lawns to remove moss and improve the vigor of grass, a process that can easily generate substantial volumes of moss. Being relatively dry and of low density, it can also be quite difficult to compost down. The key here is to mix it well with other materials. If left on its own, it may do one of two things: nothing, or turn into a slimy mess. Thankfully the scarifying process also removes some grass, which helps add those all-important Greens to kick-start decomposition. However, it is important to ensure the moss is mixed with other materials to prevent it compacting, ensure it is kept moist, and nitrogen levels are maintained.

Vegetable matter

MATERIAL TYPE (BROWN OR GREEN)	STRUCTURE (AIR CIRCULATION)	MOISTURE (CONTENT)
Green	Medium	Medium - High

Vegetable waste may be generated both in the home as a food waste (see the separate section above), but also in the garden, allotment or community garden, being made up of damaged produce, leaves, stalks and roots. Overall, they should be treated as a Green, whilst taking care not to introduce diseased materials that may contaminate the compost (see Section 4.2). Overall they should be readily compostable, although it's best to shake off any adhering soil or compost to the roots before placing in the bin.

Wood - shrub and tree trimmings

MATERIAL TYPE (BROWN OR GREEN)	STRUCTURE (AIR CIRCULATION)	MOISTURE (CONTENT)
Brown - Green	High	Low - Medium

These may be generated throughout the year, depending upon the amount of pruning and lopping required in your garden. Overall, it is likely that shrub and tree trimmings will be a mixture of woody branches and attached leaves, which means that their composition can vary depending upon the time of year. In general, you will probably need to chop up these materials first, as they are likely to be branchy and will take up a lot of room in a compost bin. Using a garden chipper or shredder can save a great deal of time; however, simply cutting branches using loppers or secateurs can be effective. The main point to remember is: the smaller the pieces, the faster they will compost down.

Wood – sawdust, shavings and chippings

MATERIAL TYPE (BROWN OR GREEN)	STRUCTURE (AIR CIRCULATION)	MOISTURE (CONTENT)
Brown	Low	Low

Sawdust, wood shavings and chippings can be added to any compost heap. They provide a good source of carbon (Brown), but do need to be balanced with some Greens and structural materials for them to compost efficiently.

It is probable that most of these materials will have been used as animal or pet bedding materials, so they will already be ready mixed with some nitrogen from the animals' manure. If you are composting cat litter made out of wood pellets, it's best to remove any feces first and follow the advice in Chapter 8, as it can contain harmful germs.

Take care not to use sawdust, shavings or chippings made out of treated wood, as wood preservatives can impair composting and contaminate the finished compost. This is discussed in more detail in the next section.

4.2 Materials to avoid

Preventing contamination

Computer scientists use the term 'garbage in, garbage out' to describe the concept that incorrect data inputs will result in poor quality or incorrect outputs. The same principle applies with a composting system: put in materials that are contaminated and the chances are that they will be there in the finished compost, reducing its usefulness and potentially causing harm to the plants and soil where is eventually used.

Contaminants are impurities that can taint compost and can be classed into three main types: physical, chemical and biological. This section looks at some of the main contaminants, highlighting what to look out for and what to avoid.

Physical contaminants

These are the things that don't rot down. Put them in a compost heap, and they'll still be there at the end in the compost. Glass, metal, aggregates and most plastics fall into this category.

It goes without saying that glass should be avoided at all costs, as broken glass can be particularly dangerous, potentially injuring anyone using the compost. It can also contaminate crops grown in the compost, and this is especially important for ready-to-eat crops (such as salads), or root vegetables, such as potatoes or carrots.

Metals should also be avoided too, as they may also cause harm to compost users in a similar way to glass. Keeping metal out of a compost bin can be a little tricky however, as many garden twines and supports contain metal. In addition, staples, nails, hooks and screws can be secretly hidden away deep inside foliage, only revealing themselves after the attached plants have decayed sufficiently. If you're composting cardboard, it may also be held together with staples.

It's best to remove any metal from the compost heap as soon as possible once spotted. Also, if you're collecting kitchen food waste for composting, please take care – I once found a tea spoon in my compost, obviously carried over from the kitchen under a mound of tea bags!

Plastic contaminants are the scourge of commercial composters, and they create problems in home composting heaps in equal measure. Garden twines, supports and plant labels are typically made of plastic, and, like metals, can be well hidden in plants. Plant pots, bags and root ball supports (such as netting) can all find their ways into the heap too. Plastic bags can be particularly difficult, as they may start to partially break down, or become brittle, during the composting process. This means that removing lots of brittle pieces of plastic in the finished compost is much harder than removing an intact plastic bag at the outset.

Wind-blown litter, such as sweet wrappers and cigarette box covers, can also inadvertently find their way into a garden, then into the compost heap, especially during fall when large quantities of fallen leaves are collected. Even if your garden isn't particularly close to a road or area where people frequent, it's surprising how much litter can be found in the aftermath of a windy day.

If you intend to compost paper or cardboard products, take care. Many of the paper wrappers we use today are laminated with a thin layer of plastic and / or foil. These can be hard to spot, and only become obvious at the end of the composting process where little plastic 'skeletons' remain visible. Tea bags and the wrappings around individualized tea bags often fall into this category. In addition, cardboard boxes may well be held together with glues, sticky tapes and staples, none of which compost; plastic delivery note pouches can also be difficult to remove.

Some plastics have been specially developed to break down during the composting process, and these are described below.

COMPOSTABLE PLASTICS

Compostable plastics are typically made out of starch or a chemical called 'polylactic acid' (PLA). Most of these products have been designed for commercial composting systems, where high temperatures are reached for long periods of time. This means that although they perform well in large-scale systems, some may struggle to degrade fast enough at the lower temperatures typically achieved in a backyard compost heap.

Identifying compostable plastics should be relatively easy: just look for a certification logo. A number of different certification schemes exist in different parts of the world, although all are based on similar principles.

North American Standard
This is based on the American Society of Testing Materials standard ASTM D6400 and is certified by the US Composting Council and the Biodegradable Products Institute.

European Standard
The European Standard is EN 13432, and there are two main certification bodies (DIN CERTCO and Vinçotte), each with their own logo: one a seedling and the other the 'OK Compost' logo.

Australasian Standard
This is AS4736-2006 and is certified by the Australasian Bioplastics Association, which covers both Australia and New Zealand. Certified products can be identified by the seedling logo.

Home Compostable Certification
Recognizing that not all products certified to the European Standard will be suitable for home composting, some certification bodies may also test these plastics at lower composting temperatures. Two such schemes exist in Europe, and one in Australasia, to a standard AS5810-2010 'Biodegradable plastics suitable for home composting'.

Caution
Plastics that are simply called 'biodegradable', 'degradable' or 'oxy-biodegradable' and are not certified to one of the standards listed above, are not suitable feedstocks for backyard compost heaps. It is unlikely that they will decompose sufficiently in a home composting environment.

Web links to sites detailing these certification schemes are listed in Chapter 9.

Chemical contaminants

In many respects, physical contaminants are relatively easy to manage: they can be seen, picked up and removed. However, this is not the case with chemical contaminants, as they cannot be seen and it may not be obvious whether or not they are present at all. They may pollute the soil to which they are applied, adversely affect plants grown in the compost, or even worse, be hazardous to human or animal health. Chemical contaminants may be naturally occurring (such as plant toxins), or man-made (such as herbicides). The main chemicals to look out for are detailed below.

Wood preservatives – Wood is treated to prevent it rotting whilst in use. However, that's exactly what we want it to do as soon as it is put in a compost heap. Treated wood may contain paints or preservatives, or contain nails, glass and other impurities. Of particular concern is wood that has been pressure treated. This often involves a chemical called chromated copper arsenate (or CCA for short), which is particularly effective in preventing the growth of wood decaying fungi. In addition, poisonous arsenic can leach out into the compost – not something any gardener would wish to happen.

Relatively recently, more environmentally friendly wood treatments have been developed that don't use CCA. However, their main aim is still to prevent wood from rotting, so these materials are best kept out of a compost heap.

Heavy metals – These are metals, such as lead, copper, cadmium and chromium. If present at high concentrations, they may adversely affect soil microbes and invertebrates, and if ingested, they may also be harmful to humans. Some heavy metals can be present in printing inks; therefore it is wise to restrict the amount of printed paper or card put in a compost heap. Research has also found that road sweepings can have high heavy metal concentrations – something to bear in mind if you plan to compost autumnal leaves swept up from busy roads.

41

Herbicides – These are widely used by gardeners to control unwanted plants, such as weeds. Most herbicides break down readily following application; however, a few can remain active for long periods of time, surviving the composting process and tainting the compost. Compost containing herbicide residues is not good for your plants!

The main herbicide to watch out for is clopyralid, which is sold in preparations used to control broadleaf weeds growing in turf, pastureland and certain crops. It is important to follow the manufacturer's instructions carefully when using these products, and ensure that any materials treated with these products are composted on their own and the compost only used on parts of the garden containing plants that are not susceptible to the herbicide.

CLOPYRALID & AMINOPYRALID HERBICIDES – TWO CAUTIONARY TALES

In the late 1990s, commercially produced compost in the USA and New Zealand was found to harm tomato plants. Following extensive laboratory testing and sleuth-like detective work, the problem was found to be caused by the herbicide clopyralid, which survived the composting process and remained active in the finished product. At the time, clopyralid was widely used in lawn care products, especially in the USA. As a result of these issues, the manufacturers withdrew some products, and restricted use of clopyralid-containing preparations. Clopyralid-containing products now contain clear guidance about composting treated plant residues.

Similar problems were experienced in 2008 by a number of British allotment holders, where manures were found to damage their plants. The problem was found to be due to the grass on which grazing animals had fed. This had been treated with products containing the herbicide aminopyralid, which was used to control weeds in pastureland. Unfortunately, the herbicide degraded slowly and passed straight through the animals' gut, remaining active in their manures. Since then, the use of aminopyralid has been restricted and stringent record-keeping procedures implemented.

And the moral of the story? Despite rigorous regulatory testing, it is difficult to predict how herbicides will behave in the environment. So, only use them with care (if at all), and always follow the manufacturer's instructions.

Plant toxins – These are produced by some plants to protect themselves from being eaten by animals; examples include the toxins in yew and ragwort. Fortunately they won't affect the composting process, or the compost itself. However, it is worth taking care if you have grazing animals and wish to use your compost in areas where they may feed.

Biological contaminants

These generally fall into two categories: weeds and disease-causing germs called pathogens.

Weeds – let's face it, they have evolved perfectly over millennia to grow and survive in our gardens. That's what they do best, and, on the whole, they have a tendency to out-compete the garden plants we actually want to grow. Naturally they are adept at surviving, even when they've been uprooted, chopped up and generally mistreated; sadly, many continue to grow despite our best attempts to ensure their demise. And the key to their survival? Their seeds, roots and rhizomes.

■ **Weed seeds** – research carried out in large-scale composting systems has shown that high temperatures can effectively destroy weed seeds. They are, however, particularly difficult to eliminate in small-scale composting systems, where sustained high temperatures are rarely, if ever, achieved. So what to do? Well, it's best to try to uproot weeds before they have set to seed. Where this is not possible, removing seed heads from the remainder of the weed and disposing of them separately is an option, albeit somewhat time-consuming. Some gardeners advocate the 'drowning method', which involves immersing weeds in water for a few weeks until the weed has completely died, then using the resultant 'liquor' as either a liquid fertilizer or booster for the compost heap. Unfortunately, I'm not aware of any research assessing the effectiveness of this method in destroying weed seeds themselves, although intuitively it seems plausible that it will reduce their viability.

■ **Weed roots and rhizomes** – on the whole, these are far easier to manage. Drying out roots or rhizomes, for example, by leaving them on a hard surface for a week or two in the sun, will, in all probability, kill the plant off. Some composters build a small platform covered in chicken wire to use as a drier, prior to composting. The downside to drying out weeds is that it can take a while to carry out effectively, especially in cooler climates, and it can take up considerable amounts of space if you happen to have a large garden.

■ **Prohibited weeds** – some weeds are so invasive (especially if they are non-native species), or are toxic to grazing animals, that governments attempt to control them. In the USA they are controlled through the Noxious Weed Act, with most states maintaining a list of noxious weeds that either prohibits or limits their distribution. In the UK, prohibited weeds include Japanese Knotweed, Giant Hogweed and Ragwort.

If you come across any of these species, then it's best to look up relevant national or local guidance, especially if they fall under any sort of legislative control. Interestingly, the British Government does permit the composting of Ragwort, with the main aim of keeping it away from grazing horses and livestock.

Diseased plants – these are generally caused by microorganisms, such as fungi, bacteria and viruses. Examples include brassica clubroot, onion white rot and potato blight; the latter is the disease that led to the infamous nineteenth century famine across Ireland. Again, there has been a significant amount of scientific research investigating the effects of large-scale composting on the viability of these pathogens. Sustained high temperatures in well-managed systems have been shown to eradicate pathogens of concern; however, this cannot be extended to smaller-scale low-temperature compost piles. As a rule of thumb, it's best to keep diseased plants out of a garden compost heap to prevent its reintroduction back into the soil.

Plant pests – these include an array of insects that either eat the plant directly or lay their eggs on the plant, only to emerge and eat it at a later date. Examples include, but are not limited to: aphids, ants, flies, caterpillars, moths and mites. Generally, if the plant dies, then there will be no food for the insect to eat, so they will die too. Putting these in a compost heap should not pose any significant problem. However, some pests do eat dead or decaying vegetation, so try to avoid composting these if at all possible. If you are uncertain, then it's best to err on the side of caution and dispose of these materials elsewhere.

MATERIALS TO AVOID AT ALL COSTS

▓ **Glass** – it won't rot down, and may well break into tiny pieces that can potentially injure people and animals.

▓ **Plastic** – unless it has been specifically certified as 'compostable', then avoid it. Traditional plastic films made of polyethylene will breakdown and decompose eventually. However, this may take many decades, and in the meantime, it will break up into tiny fragments that can harm animals.

▓ **Nappies / diapers** – although some are marketed as being 'biodegradable' they do not degrade sufficiently in a home composting system due to the lower temperatures.

▓ **Cigarette butts** – the filters used in cigarettes are made of a polymer that will only rot down very slowly.

▓ **Treated wood** – the chemicals are designed to prevent the wood rotting. They may be toxic to humans and can leach out into the soil. Paints, laminates and varnishes are also undesirable.

▓ **Invasive species** – these need sustained high temperatures to kill them off. Composting runs the risk of re-introducing them back into your garden.

MATERIALS TO USE WITH CAUTION

There are a number of materials that can be composted, but may contaminate the compost or be harmful to people or plants under certain circumstances. Whether you choose to put them in your compost heap will depend upon how you manage your system and how you plan to use the compost.

▩ **Weeds with seeds** – the seeds may remain viable during the composting process, only to germinate and grow into new weeds.

▩ **Diseased plants or produce** – the disease-causing organisms may remain viable during the composting process, so may contaminate soil to which the compost has been applied.

▩ **Pest-infested plants** – some of these may survive the composting process to re-infect your garden plants.

▩ **Herbicide-treated plants** – some herbicides (clopyralid, in particular) may remain active in the compost and have the potential to harm plants to which compost has been applied. Always read the label and follow the manufacturers' instructions.

▩ **Cat litter** – some litters are made out of pelleted sawdust (wood) or straw, which can be composted. However, take care to remove any feces (poop) beforehand, and be aware that the litter may still contain harmful germs.

▩ **Paper and card** – these may be laminated with plastic film, contain staples, tapes and packaging pouches which don't decompose. In addition the glosses and inks used in some prints may contaminate the compost.

▩ **Compostable plastics** – plastics that have been certified as compostable are suitable for large-scale composting systems, although they may take longer to breakdown in a garden compost heap.

▩ **Street sweepings** – these may contain litter, dog feces, pollutants from vehicles and grit. It's best to know where your sweepings are from before composting.

▩ **Compostable diaper / nappy liners** – some of these are made out of paper, so will compost easily. However, take care, as they may contain harmful germs, and may also smell.

▩ **Vacuum cleaner contents** – these can contain synthetic carpet fibers and plastic fragments (especially if there are children in the house) that won't decompose. In addition, anti-stain protectors may also be present on carpet fibers. In general, vacuum cleaner contents don't tend to absorb moisture very well, so they remain as an intact mass within the compost bin.

4.3 Mix ratios

Over the years, I've been asked by lots of different people what the best composting recipes are. Fundamentally, people want to know what they should mix together and in what proportions, which seems sensible enough. However, this presents a number of different challenges, which makes it decidedly difficult to be prescriptive.

Recipes, by definition, specify what should be used (the ingredients), in what proportions they should be mixed (take two eggs, a knob of butter and three potatoes, for example), and how they should be prepared and cooked (chop, mix then cook at such-and-such temperature for so many minutes). With composting, it is far harder to define the ingredients, measure how much there is and control how it will be processed.

The nature of gardening means that we need to be flexible and respond to the needs of our garden: managing grass clippings in summer, fallen leaves in fall and tree prunings in the winter. The emergent nature of the materials we compost, coupled with the fact that they will begin to decay as soon as they have been harvested, means that we need to compost them pretty much straight away. Put simply: we have very little control over our ingredients.

Secondly, recipes define proportions, usually on a weight basis. However, this is nigh on impossible in a garden setting where the density and moisture content of different feedstocks can vary considerably, and measuring large volumes of materials can be challenging, to say the least. If you're considering using blending ratios, then it may be a good idea to think about a bucketful or wheelbarrow load.

And finally, processing. Unlike a kitchen where the cooking conditions can be carefully controlled, compost bins are sited outdoors and are subject to ever-changing weather, notably extremes in temperature, sunshine and precipitation. These all conspire to make predicting how the composting process will progress somewhat difficult.

Not convinced and still want a recipe? OK then, a useful rule of thumb is to mix the Greens and Browns in a ratio of 2:1. I'll leave it up to you to define what you mean by a 'part'!

MIXING 'RULE OF THUMB'
Mix 2 Parts Green to 1 Part Brown.

4.4 Don't have a decent mix of materials?

Throughout this book, I've explained the importance of ensuring that a compost heap is made up of a decent mix of Greens and Browns. The configuration of some gardens, however, means that it is nigh on impossible to obtain a steady ratio of Greens and Browns throughout the year. Instead, gardeners with large lawns, for example, are likely to generate large quantities of grass clippings with precious little else to mix it with. Others may be inundated with leaves during the autumn and winter months. It is at times like this that a contingency plan is needed to prevent problems arising and to allow composting to take place as best it can. Here are some suggestions.

A surplus of Greens

These are most likely to be grass clippings, which, on their own, will rapidly start to decompose, turning into a smelly, slimy mess. This problem is particularly acute in plastic compost bins where air circulation can be poor and moisture is slow to evaporate.

The best way to prevent this happening is to reduce the moisture content of the cut grass, mix it with some Browns (to balance the carbon-to-nitrogen ratio) and provide it with some structural material. Where it is not possible to do all three at once, then try the following:

■ Partially dry the grass out by stacking it in long rows for a few days. By keeping the height of the rows relatively low (no more than 1 foot or 30 cm high) and maximizing its surface area, this will reduce the moisture content and, hence, its potential to turn odorous. (Incidentally, this is the technique formerly used by farmers to dry out hay and straw before large scale mechanization took over. The term 'windrow' has now been adopted by the composting industry to describe rows of composting materials; hence windrow composting!).

■ Mix the grass into an existing compost heap. This may act as a booster for the heap as well as adding additional moisture. Either way, by doing this you will be adding some structure to the grass which will help prevent it slumping and turning smelly.

■ Keep a stockpile of 'Browns' on hand especially for this purpose. Autumnal leaves are an easy and simple way of achieving this, as are sticks and branches. Having a supply of carbon-rich materials on hand means that problems can be easily and effectively averted. However, if you have too many Browns, then read below.

■ Add some shredded or scrunched up paper to the grass clippings. This will help absorb water, provide structure and add carbon. However, take care to ensure that the paper does not introduce any unwanted contaminants through printing inks or laminates.

■ If any of the above are not possible, or the amount of grass remains too large, then send it off for recycling using a municipal green waste collection service. This may be via a curbside collection scheme, or centralized drop-off point, such as at a local household waste recycling center. Either way, never send them to landfill in the black (residual waste) bin. Alternatively, neighbors or a local community composting group may be able to help out.

A surplus of Browns

As Browns, by their very nature, tend to decompose at a slow rate, this means that they have the potential to build up and start to clutter a garden. Keeping a small stockpile of Browns on-hand to mix with Greens during the spring and summer months is sensible; however, when this stockpile becomes too large, alternative measures are called for. Here are a few suggestions:

■ Reduce the size of logs and branches by cutting them up using a pair of loppers or a saw. The smaller flexible branches may then be easier to compact and fit into a compost bin. The larger branches and small logs can be stacked to create a wildlife habitat, encouraging invertebrates and hibernating animals to take up residence.

- Invest in a garden chipper or shredder if funds and space permit. Electric shredders can work well with small branches, but can struggle with larger, harder sticks. Alternatively enquire if there is a mobile shredding service in place locally.

- Recycle it through a municipal composting scheme; your municipality may offer a collection service from your home, or provide a centralized collection point at a recycling center.

- Enquire to see if there is a local community composting group or master composter, who may be able to take the material or provide assistance.

- Gather autumnal leaves and stack in a dedicated bin to make leaf mold. There's an explanation of how to do this in Section 6.1.

- Cut and use the larger branches as a biomass fuel, if you are lucky enough to have a wood burning stove installed in your house.

- Finally, as a last resort, the woody material may need to be burned in the garden. Environmentally, this is not the best option, and may be subject to local air quality controls, so it's best to enquire with your municipality first. Remember the ash is a useful fertilizer, as it's high in potassium.

4.5 Summary

This chapter has focused on preventing problems occurring in the first place. We've looked at the properties of the main types of materials that can be composted, and some of the things you may need to keep an eye on. We've also looked at some of the things that can cause problems and potentially contaminate your compost. Finally, we've discussed mixing ratios and I've made a few suggestions as to how you can practically manage a surplus of either Greens or Browns.

5 Troubleshooting

Now that we've discussed the key elements needed for successful composting, we can turn our attention to systems that don't quite go to plan. Thankfully, there are few things that can go 'really' wrong with a small, garden compost heap. Chances are that most of the time the process has just slowed down and is decomposing in a less than ideal way. In most cases, making a few simple adjustments can make a huge difference to the way the materials compost.

This chapter looks at some of the main problems that can arise, then makes practical suggestions as to how you can tweak the composting materials to improve the process. I've introduced a simple framework to assess and address problems, then described some of the problems that may be encountered with the composting process, unwanted visitors and the compost itself.

However, before we delve into the nitty gritty of troubleshooting, it's helpful to know what to expect out of a backyard compost heap; that is, how long it should take and what it should look like.

5.1 Managing expectations

Garden centers, websites and promotional literature selling compost bins often depict nice new shiny bins, with garden waste visible at the top of the bin and nice, dark, uniform compost seeping out of the little door at the bottom. The implication is that the 'stuff' at the top will miraculously transform itself into the 'stuff' at the bottom within a few months. It can't and it won't, so please don't expect it to.

Typically, it can take anywhere between one and three years for garden waste materials to decompose sufficiently so that the end product can be called 'compost'. If you add materials to the top of your heap in a piecemeal fashion, then clearly the older materials at the bottom are more likely to have decomposed to a far greater extent than the newer materials at the top. In this case, you'll need to harvest the bottom few layers little by little.

Similarly, don't expect your garden compost to look, feel or perform like the 'compost' you may buy in bags from the garden center or home improvement (DIY) store. This commercially manufactured compost has been carefully manufactured and blended to create a technical product designed for a very specific end use. Some products may not even contain compost at all, and may instead be a blend of other materials, such as peat, coir (coconut fiber) or bark. Homemade compost is very different to this, and will, in all certainty, contain a mixture of well-rotted materials mixed with partially decomposed plants, such as twigs or branches.

Please bear this in mind when assessing the performance of your compost heap. If you're convinced you still have a problem, then read on.

5.2 Diagnosing the problem

Problem-solving generally involves a minimum of four steps, which I've depicted as a cycle shown below.

The problem-solving cycle

In general, when considering backyard composting, problems tend to center around three main areas:

■ **The composting process** – the materials don't appear to be decomposing efficiently.

■ **Unwanted guests** – creatures that might decide to set up home in our compost bin.

■ **The compost itself** – it doesn't look, feel or behave how you expect it should.

Most troubleshooting will involve adjusting one or more of the three main components needed for effective composting. If you've managed to read Chapter 3, which sets out the composting basics, then hopefully it should be clear that a decent mix of materials is needed (a balance of Greens and Browns); they need to have an adequate structure to allow gases to circulate within the pile; and they need to be moist, but not too wet. Where unwanted visitors are concerned, other additional measures may also be needed.

Although they may not always be apparent immediately, making simple changes can help get things back on course. Keep on checking though, as more than one intervention may be required depending upon the severity of the problem.

5.3 Troubleshooting common problems

The composting process

My compost heap doesn't heat up

Chances are that the temperature of the materials in your compost heap is at least a few degrees above ambient, although it may not be noticeable. Unlike large-scale commercial composting, where high temperatures can be attained for long periods of time, small-scale systems simply lack the thermal mass needed to heat up sufficiently. Because garden compost bins are generally less than 1 yd³ (about 1 m³) in volume (and some of the plastic bought varieties are often about a third of this), this means that they have a relatively large surface area compared to their volume through which to lose heat. As a result, the heat generated by the composting microbes will be rapidly lost through the sides of the bin, rather than being retained in the heap to increase the temperature.

Low composting temperatures may also be a result of a poor mix of materials, especially if there is a surplus of Browns and insufficient Greens, or if there is insufficient moisture to allow the composting microbes to decompose the wastes. If the structure of the materials is too branchy, this may mean that there will be too much air circulating, removing valuable heat and allowing too much water to evaporate. Finally, seasonal changes in ambient temperature and wind chill may also play a role, especially during cold winters.

If you want to try to increase the temperature of your compost heap, you could try one or more of the following:

- **Increase the size of your heap** – if your compost heap is of the homemade variety, then try to consolidate a few and create one much larger heap. Try to aim for a minimum of 1 yd3 (which is slightly smaller than 1 m3). The larger the volume, the greater the thermal mass and the lower the surface area to volume ratio. Obviously, a cube will be more effective at retaining heat than a long, thin rectangle.
- **Insulate your heap** – you could try to reduce the rate at which heat is lost from your heap by covering the top and sides of the bin with an old carpet, or by buying an insulated compost bin. Either way, remember that the more insulation you place around your heap, the less freely air will be able to circulate.
- **Adjust the mix of materials** – adding additional Greens to your composting bin will act as a 'booster' and help kick-start the composting microbes. Grass clippings are a very useful way of adding valuable nitrogen and moisture to the heap, although remember to mix them well with the other materials so that they don't compact.
- **Turn your heap** – it might be that the center of your composting heap has become depleted of oxygen and a good mix up is called for. Removing the compost from its bin and 'turning' it will help mix the materials, bring in fresh air and allowing stale air to escape.

- **Compact the materials if they are particularly branchy** – if you have a lot of woody, branchy Browns in your heap, it might be that they are allowing too much air to circulate, causing valuable heat energy to dissipate. Reducing their size, for example by compacting them with a spade, or even jumping on them, may help. However, please take care and do it safely.
- **Move your heap** – relocating your heap to a warmer part of the garden could also help. Placing the bin in a sunny area which is less prone to frosts in winter and one which is sheltered from prevailing winds, should reduce the rate of heat loss. However, sunny spots may also unduly dry the materials out during the summer months, so this is worth bearing in mind.
- **Apply a compost booster / accelerator** – commercial preparations are available that purport to add enzymes and microbial mixes that can kick-start composting.

It's not steaming

And, quite frankly, it is unlikely that it will! Steam is only visible on cold days when water vapor released from the heap condenses into tiny droplets. Steam isn't an essential indicator of a well-run compost pile, so why worry?

If you are concerned about your heap not heating up, then follow the advice in the previous section.

It's taking too long to compost

As we've discussed previously, it's important to adjust your expectations to what is reasonably practicable given the mix of materials you have available, the prevailing weather conditions in your backyard or garden, and the time you can afford to spend managing your compost. Some composting websites suggest that it is possible to make compost in as little as three months; however, in the vast majority of cases, this expectation would be unrealistic.

Let's make a comparison with commercial composting processes, which are actively managed and reach high composting temperatures. These systems generally take a minimum of eight to twelve weeks to produce a stable, mature compost – anything less and the compost is likely to smell and be harmful to plants. It would therefore be unreasonable to assume that lower temperature, passively managed systems could create compost in a shorter timeframe. As such, unless the compost is being made out of easy-to-degrade materials (such as food wastes) in an actively managed heap, then it is highly unlikely that useable compost will be available within three months. Please don't plan on anything less.

(The possible exception to this is Bokashi composting, or when proprietary accelerators have been used, in which case it's best to follow the manufacturer's instructions. Bokashi composting is described in Section 6.3).

How long should it take?

There are basically two main phases to the composting process: the first is the rotting down of the materials to provide food for the microbes; and the second, is the formation of humus (a process termed 'humification'). Humus is the stuff that makes compost look dark, and is responsible for the organic matter in our soils. This step is the one that takes the longest, especially if there are lots of woody Browns in the mix. The rates at which degradation and humification can occur therefore dictate how long it takes to produce compost.

Overall, it would be reasonable to expect to harvest your compost after about a year. The length of time will, of course, depend upon the materials you've composted, the extent to which the process has been managed, and the weather. Rather than stick to pre-defined timescales, it's probably best to assess the compost using the following simple criteria:

- Is it dark?
- Does it have a pleasant earthy smell?
- Other than twigs, is it reasonably crumbly?

If the answer to all three questions is 'yes', then it's probably ready to use. Take care though. It's quite probable that the compost will still contain some twigs and woody material that haven't yet fully degraded, so these may need to be picked or sieved out, depending upon how your compost will be used.

Indicative composting timescales are shown in the box below.

TYPICAL COMPOSTING TIMESCALES

▓ **Six months to one year** – food wastes, grass clippings and other easy-to-degrade Greens.

▓ **One to two years** – most Greens and some Browns including sappy bedding plants, autumnal leaves and small twigs.

▓ **Two to three years** – most Browns, including cut branches and straw.

▓ **Longer than three years** – dried wood, including larger branches and tree trunks.

In general, expect it to take anywhere between **one and three years** to make garden compost.

If you wish to reduce the length of time it takes to make compost then you'll need to make sure that the heap has:

■ A decent mix of Greens and Browns;
■ Adequate moisture levels, and
■ Sufficient air.

It's too dry

Unfortunately this is a common problem, especially with wooden compost bins. In particular, it can be hard to prevent the outer layers from drying out, especially in dry climates or during summer months if the bin is sited in a sunny position. As we've discovered, dry materials don't degrade as there is insufficient water on the surface for the microbes to do their job, so the aim must be to increase the moisture content.

Common sense would tell us that if it's dry, add some water. Unfortunately the majority of this simply runs straight off into the ground and any remaining water soon disappears. The best way to address a dry compost heap is to reduce the amount of water that will evaporate from it. Some of the tips suggested to increase temperature will also help, such as:

- Reducing the surface area to volume ratio of the bin, by grouping a few heaps together.
- Covering the top and sides with cardboard, old carpet or a custom-made lid.
- Moving the heap to a part of the garden that isn't in direct sunlight or prone to prevailing winds.
- Adding wet Greens, such as freshly cut grass clippings – these tend to retain moisture and can kick-start the composting process.

Again, please bear in mind that by reducing evaporation, you will also be reducing the amount of air that can diffuse into the bin, so it may have the opposite effect and start to smell.

It's too wet

This tends to happen when a lot of Greens are put into the bin, especially if the bin is made out of plastic. Materials such as grass clippings and vegetable peelings contain a lot of moisture and they also tend to compact easily. This has two effects: it means that air doesn't circulate particularly well within the composting heap, which then reduces evaporation. Water may also drain down to the bottom layers of compost, which, in turn, can start to smell.

The best way to overcome a soggy compost heap is to open up the structure of the material and add drier materials. First of all, take the compost out of the bin and try to let it dry off in the air for a while (assuming it's not raining, of course). If you have Browns at hand, then mixing these with the compost can help restore the balance.

If your garden is small and you are composting a lot of vegetable peelings from your kitchen or lawn mowings, then it may be easier said than done to find surplus Browns. In which case, shredded paper is a great way of absorbing surplus water and adding much-needed structure. Take care not to use too much glossy or color printed paper though.

If you're still experience problems, then consider building up a layer of twigs or other woody materials at the bottom of the heap. This will help improve drainage of surplus moisture into the soil.

Finally, if you still have soggy compost and you use a plastic bin, then drill some holes into the sides. I've resorted to this before, and it certainly helps, but is no substitution for a well-balanced mix of materials.

My compost smells

This problem is prone to happen in wet, soggy compost. Grass clippings, in particular, are notorious for smelling, especially if left in a warm spot for a few days. This is because grass provides a readily available, nutritious source of food for our composting microbes. They become very active very quickly, using up any available oxygen present and releasing heat energy. As a result, the grass starts to warm up and compact even further, meaning that there is little structure for fresh air to circulate. This chain of events conspires to create conditions ideal for the growth of the oxygen-hating (anaerobic) bacteria. As we discussed previously in Chapter 3, these microbes often release odorous by-products, which are the root cause of the problem.

Typical odors include:

- **A rotten egg smell** – this is caused by hydrogen sulfide, and is typical of wet, oxygen-depleted environments. It means that there is no air circulating into this part of the composting mass.
- **Sweet sickly smells** – these are caused by organic acids released when grass begins to decay in the absence of oxygen. This is the process used by farmers to create silage.
- **Ammonia** – this is caused when excess nitrogen is released into the air, and means that there are too many Greens and insufficient Browns to balance the carbon-to-nitrogen ratio. It is particularly noticeable if you are composting manures.

Should you encounter any of these problems, then it is best to follow the advice suggested for overcoming soggy compost, as the two symptoms usually go hand-in-hand.

To recap, they include:

- Drying the compost out.
- Adding structural materials, such as shredded paper, to open up the air pockets.
- Adding additional Browns to balance any surplus nitrogen.
- Creating a layer of branches and twigs at the bottom of the bin to improve drainage.
- Drilling holes in the bottom of the bin.

If it really, really does smell, then you'll need to resort to a damage limitation exercise. In practice, there is very little that can be done with very smelly compost. Grass that has started to turn to silage can be troublesome as the organic acids can smell for a very long period of time. (They also stick to clothes and skin and are difficult to wash out – see my 'Salutary Lesson' below). If this happens, it is best to try to prevent the odors being released into the air by covering the compost with a layer of either soil or mature compost. The layer need only be about 2 inches (5 cm) deep. It will act as a simple biofilter, helping to neutralize any gases that are released, and will help add back some of the essential oxygen-loving microbes. It's best to leave the heap for a good few months and let nature take its course.

A SALUTARY LESSON

Many years ago, when I first started composting, I proudly made a wire mesh bin at the top of the garden. Looking for feedstocks, I approached my neighbors and managed to source some grass clippings. "Great", I thought. The only problem was that they had been left to fester in black plastic bin bags for well over a week. "They're a bit smelly", my neighbor exclaimed. "No problem", I thought, "let the air get to them and they'll be fine".

How wrong was I? These grass clipping stank terribly for weeks. Not only did they smell at a distance, but if they came into contact with skin, then the smell transferred across. It was very hard to wash off.

The lesson? Don't try to compost very smelly materials, especially if they have been stored in the absence of air for some time, as the products of oxygen-free decomposition are odorous.

Plants are growing into my compost bin

It has been known for some plants to migrate through the soil towards, and into, a compost heap, harvesting any nutrients that may leach out of the composting mass. Bamboo, in particular, has been known to do this (so-called running bamboo). Once in the soil, and without resorting to herbicide treatment, it can be difficult to control. If this does happen, there are a number of options available to try to address it:

■ Move the compost bin to a different part of the garden away from the invasive plant.

■ Place a sheet or two of a fine-mesh chicken wire at the bottom of the bin; it may be necessary to fold over a few sheets to reduce the mesh size sufficiently. This won't entirely prevent the plant growing into the bin, but it will certainly make it harder for them, and easier for you to remove them when you harvest your compost.

■ Use a tumbler compost bin, in which the composting materials are held above ground and don't come into contact with the soil.

Do I need to add worms?

Unless you specifically plan to make vermicompost, then, no, you'll not need to add worms. As long as there is a reasonable mix of materials, the worms already living in your garden will naturally colonize the bin and help with the degradation process.

Should I use an activator?

Some people swear by activators and they can be particularly helpful if you struggle to source a balanced mix of Green and Browns in your garden. However, all things being equal, if you follow the advice in this book and achieve a balance of materials, water and air, then an activator is not essential.

Unwanted visitors

Rodents

The pitter-patter of tiny feet in any compost heap is never welcome. Believe it or not, rodents, such as rats and mice, are, in many ways similar to us humans: we all like somewhere nice, warm and secure to live and bring up our offspring, and we also need a ready source of good, healthy food to eat. And that's the problem, compost heaps can provide some, if not all of these things for a local rodent population.

I have only ever experienced problems with rats in my bin on two occasions. Once during a particularly harsh winter, when we endured many weeks of sub-zero temperatures and heavy snowfall; and once during a very wet winter, when the ground was saturated with water. On both occasions my compost bin provided the local rats with a relatively warm and dry place to live. Fortunately, as soon as the weather improved, they soon scarpered.

To prevent rodents taking up residency in the first place, it's best to avoid putting meat, fish or dairy products in your compost heap. These provide a gourmet meal and will certainly attract any hungry animal. Dry compost heaps can also provide a ready-made bed. If you have mice, then check if the compost is too dry, and follow the advice about increasing the moisture content.

Rats are also creatures of habit, with generations following the same 'rat runs'. They also dig underground tunnels, so simply removing a food source or making their beds a bit damper will not necessarily deter them. Once a rat has taken the time and effort to dig a tunnel into the bottom of a compost bin, then it's not going to forsake it without a struggle. Should this happen, then more drastic action may well be needed.

If possible, the simplest and most effective way is to pick up and relocate the compost bin to a different part of the garden. Should this not be possible, or if the rats manage to find their newly-sited home, then physically preventing them gaining access is the next option. The easiest way to achieve this is to wrap some chicken wire over the bottom of the bin. Bearing in mind rats' tunneling prowess, this may need to be made of heavy-duty narrow-gauge mesh. You may also need to secure it to the soil with some tent pegs or similar, and ensure that the bin is securely attached.

As rats are eminently adaptable, intelligent creatures, a combination of all three measures may be required: removing their food, making their home less hospitable and impeding access.

On a final note, I have found that since acquiring a cat and composting the spent litter (which is made of wood and straw pellets), this has had the effect of discouraging rodents. Thankfully, the smell of cat urine is wholly unattractive to them. If you do this, please remember *never* to put cat feces (poop) on your compost heap, as this can carry disease-causing germs. This should be disposed of separately, either down the sewer or in the waste bin.

Raccoons, chipmunks and squirrels

Depending upon where you live, a compost heap containing meat, fish and /or cooked food, can also attract larger mammals, such as raccoons, chipmunks, squirrels, opossums and even bears. These scavengers are opportunists, and will sniff out a good meal at the first opportunity. So, the easiest way to deter them is not to put cooked foods, meat, fish or dairy products into your compost heap in the first place.

Flies, ants and other creepy crawlies

The presence of insects and other invertebrates in a composting heap is natural, with many helping to mix the materials and feed off the composting microbes; these are the secondary and tertiary consumers described in Chapter 3. Although the vast majority are beneficial, some can become a nuisance at certain times of the year. Late summer and early fall seems to be the time when flies and ants take up residency in large numbers.

The chances are, if you have lots of small flies in your bin, these will be fruit flies. As their name suggests, they tend to be attracted to fruit, and can be troublesome if you added any sizeable quantity – damaged windfall and the leftovers after jam making are particular attractants. To reduce this, try to cover any fruit with a barrier, such as shredded newspaper, or even a thin layer of soil. Physically preventing the flies from reaching the fruit is the simplest means of controlling them. If this doesn't work, then it's probably best to leave the lid off the bin. This will help dry the fruit out (making it less attractive to the flies) and expose them to the local bird population.

Either way, as soon as the fruit has degraded beyond a certain point, they will up-sticks and move on.

Ants, on the other hand, may well decide that your compost heap is a good place to build a home and lay their eggs. This isn't a problem in itself, as they can be beneficial to the composting process. However, if ants aren't to your liking, then try upping the moisture levels, as they tend to prefer dry heaps.

Unlike ants, woodlice tend to prefer damp conditions, where they feed on decaying vegetation. They are an integral part of the complex food chain that typifies a healthy compost heap and shouldn't be discouraged. However, if you don't wish to see them, simply leave the lid off. Woodlice don't like light, so they will soon seek out shady places to live where they are safe from scavenging birds and where they won't dry out.

Snakes

Snakes have been known to take up residence in a compost heap, where they may use it as a safe place to warm up or lay their eggs. Depending upon where you live, and the type of snake involved, this may be inconsequential, or potentially dangerous. It's always best to seek expert advice if you are in doubt. Either way, it's best to leave the snakes alone and let them leave of their own accord.

In Northern Europe, if you're lucky, you may find that a grass snake has decided to use your compost heap to lay its eggs, especially if you live near a pond or river. The eggs are generally laid in June or July and will take up to 10 weeks to hatch. Grass snakes are non-venomous and will not harm people, preferring instead to eat frogs, toads and worms.

Hibernating animals

If your compost heap is particularly dry and sheltered, you may find that an animal or two has decided to overwinter. Examples include frogs, toads, newts and hedgehogs. If this happens, it's best to leave them until they emerge again in the spring.

In some areas of Europe, hedgehog numbers are in decline, so by providing them with a safe place to overwinter, you will be helping these little mammals who are so adept at keeping the slugs and caterpillars at bay during the summer.

The compost itself

Homemade garden compost is very different to the 'multi-purpose compost' sold in bags. Don't expect your own compost to look, feel or behave like the bagged variety. It's worth bearing in mind that the marketing photos depicting nice shiny compost bins with fine, crumbly compost oozing out of the bottom are just that: marketing photos. I've tried to summarize some of the main concerns people sometimes have about their compost below.

It hasn't composted – it's still got big bits in it

Don't panic! Your garden compost may well look rough, containing twigs, branches and other pieces of woody material that haven't yet fully decomposed. That's fine. Remember, if your compost is dark and has a pleasant earthy smell, then it's probably ready to use.

Depending on how you wish to use your compost, you may first want to sieve it to remove these larger pieces. This can be achieved simply by picking out large fragments by hand, or by using a hand-held sieve; however, if you have any sizeable quantity, this may well be a time-consuming and back-breaking exercise. Instead it's worth considering either making or buying a larger sieve.

Compost sieves, 'riddles' or 'screens' as they are sometimes called, can range in complexity. At their simplest they can be made out of a rigid wire mesh that can be laid over a wheelbarrow or propped up on a frame. The mesh size will obviously dictate how fine the screened compost will be. More complex varieties involve a rotating drum, which can be either turned by hand or, for the more adventurous, coupled to a small motor. Some examples are shown below.

Handheld sieve	Static frame sieve	Mechanical rotating sieve

Different types of compost sieves

Irrespective of the type of sieve you use, the compost shouldn't be too wet, otherwise it will tend to form large clumps and won't easily fall through the screen.

I can't get the compost out of my bin

If you've bought a plastic bin with a little door at the bottom and expect to be able to easily remove your finished compost through it, then think again. This is yet another marketing gimmick and is unlikely to work in practice. As it has decomposed, compost becomes more compact (that is, denser) than the materials it was made from. Coupled with the weight of the Greens and Browns pressing down from above it, this means that it's difficult to insert a garden fork or trowel and extract the compost. Instead, the easiest way is to remove the bin, harvest what you need, then replace the feedstocks back inside. This can be achieved quite simply with a conical plastic bin, but may involve an element of deconstructing a larger wooden bin (such as removing the front panels).

Parts of my heap have composted and others haven't

This can happen if you have a layer of Browns that haven't been adequately mixed with the Greens, or if parts of the heap have dried out, especially around the edges. If this has occurred, then simply harvest the parts that have composted and put the remainder back into the heap, trying to mix in some more Greens and increase the moisture levels.

If you have a small garden and only one active compost bin, then you may wish to use the bottom layers of compost, whilst the top layers are still un-degraded. That's fine, simply harvest what you require and shovel the remaining partially composted materials back into the bin, giving them a good mix as you do so.

There are weeds growing in the compost

Controlling the spread of weeds is an ongoing battle in most low-temperature composting systems. By being careful not to put living weeds and their seed heads into the heap in the first place, this problem can be reduced. However, despite our best efforts some weeds still manage to survive and flourish. If they do, there are a number of different options open to you.

- **Remove the weeds** – large weeds can be removed from the compost by hand. By allowing them to dry out fully before putting them back onto the heap, this should reduce the chances of them continuing to cause problems. Alternatively, they can be 'drowned' in a bucket of water, or placed inside a black plastic bag for a few weeks to kill them off.

- **Dig the compost into the soil** – if you plan to use the compost as a soil conditioner or planting medium, rather than as a surface mulch, then it's best to dig weedy compost into the soil, preferably in a shady spot that doesn't receive too much light. This might not be ideal, but it will help impede their growth and reduce the chances of them spreading.

- **Turn the compost** – if you decide that there are too many weeds for your liking, then it's best to try to break up any roots as far as reasonably practicable, then 'turn' (mix) the compost in with itself and leave it to compost for a while longer. This should hopefully kill off some of the weeds and weaken any remaining ones. Remember that weeds need light to grow, so by covering your compost heap and keeping out as much light as possible, this will help keep them at bay.

Occasionally, other unwanted plants may grow out of kitchen scraps, including tomatoes, potatoes and avocadoes. Tomatoes, in particular, produce very resistant seeds that have been known to pass through the human digestive tract, survive the journey through the sewers and the rigors of the waste water treatment works. No wonder they sometimes appear to sprout up in a compost heap!

Discarded potatoes and avocado stones can also remain viable and show signs of growth. A well-aimed garden spade should be sufficient to break them up and help prevent further growth. Ideally, let them dry out fully before returning to the heap.

The compost's gone 'dusty'

Sometimes a fine white 'dust' may appear on the surface of compost; this is nothing to be alarmed about and is perfectly natural. It is made up of the spores of some of the composting microbes and is sometimes called 'fire fang'. It tends to form on feedstocks, particularly grass clippings, which have heated up to high temperatures, then cooled down.

Fire fang won't harm your compost or your plants, but take care not to breathe it in. If you need to move compost that shows signs of fire fang, then it's good practice to dampen it down with some water first to reduce the likelihood of dispersing it into the air. Further advice on health and safety is given in Chapter 8.

The compost's killed my plants

Oh dear, that's not good. The last thing we need is for your lovingly made compost to damage your plants. In practice, it's very unlikely that this will happen, but it is possible, especially if you are using the compost as a growing medium (potting soil) for containerized plants. Possible reasons for damaging your plants include the following:

- **The compost is too 'fresh'** – it might be that the compost is still immature and some of the by-products that form during the degradation process haven't been fully degraded themselves. Notably, these include the organic acids that can form if the heap has started to decompose in the absence of oxygen, or ammonia if there is a lot of nitrogen-rich material present, such as manure. If this happens, then it's best to 'turn' the compost and allow it to continue to mature for a few more months as a minimum.
- **The compost is still 'unstable'** – this means that it is still actively degrading. As the composting microbes are very adept at scavenging whatever nutrients they need, if there is still a surplus of carbon to be broken down, they will extract whatever nitrogen they need from the soil so that they can keep on 'eating'. This then has a knock-on effect and deprives the plants of the nitrogen they need to grow. It is called 'nitrogen lock-up' and effectively immobilizes any free nitrogen in the soil. Allow the compost to decompose for a few more months before using it.
- **It contains an herbicide** – if you have treated some of your plants with a proprietary herbicide and then put them in your compost bin, it is possible that residues remain. Most notable is the herbicide clopyralid, which can remain active for long periods of time (see Section 4.2). It is always advisable to follow the manufacturer's instructions if you need to use an herbicide. If you suspect your compost may be tainted, then then it's best to keep it away from susceptible plants and compost it for at least another year. Research shows that herbicides do degrade over time, although some take longer than others.

The compost's full of creepy crawlies

These creepy crawly invertebrates are an essential part of the composting process, but may not be desirable if you wish to screen your compost or use it in pots in a greenhouse or conservatory. If you feel you have too many of these creepy crawlies for your liking, then the easiest way to get rid of them is to remove the compost from its bin and spread it out on the ground for a few hours. Most of them dislike daylight and will try to hide away to prevent being eaten and drying out. In addition, your local bird population will be given a gourmet meal and will readily take advantage of any insect offerings you can provide.

5.4 Summary

This chapter has explored some of the difficulties that any composter, whether novice or experienced, can encounter. As we've seen, most composting problems are a direct result of too little or too much of one of the vital composting components. Hopefully, by having read this chapter, you can now diagnose and put right most problems you may come across.

6 Alternative Composting Methods

The previous chapters have focused on how to create a balanced mix of materials for your compost heap – one that has a decent blend of Browns and Greens, sufficient air and enough moisture. However, there are a few alternatives that seem to buck these basic requirements. So what's going on?

Although all composting processes rely on the same basic principles, the ways in which the processes are managed can differ. The five main alternatives to traditional garden composting are described in this chapter.

6.1 Leaf mold composting

Fallen leaves are a valuable source of organic matter, but they are also low in nutrients. Gathering leaves and piling them up on their own doesn't, at first glance, seem to be a sensible thing to do. Surely they'll need to be mixed with some Greens?

Not necessarily. Unlike most other composting materials, deciduous leaves have a very high surface area to volume ratio, as they've evolved to capture light from the sun. In turn, this means that there is a large surface for the composting microbes to work on, so decomposition is speeded up compared with bulkier Browns (such as twigs and branches). Adding Greens will, of course, speed up the process, but this may curtail the usefulness of the product. The high organic matter content and relatively fine structure of leaf mold compost makes it particularly suitable for low-nutrient applications, such as in potting mixes (see Chapter 7).

The key to efficient leaf mold composting is to break the leaves up as much as possible (to maximize their surface area) and to keep them moist. Leaves can be composted in a normal compost bin, but they need to be kept damp. Alternatively, you could manufacture a simple bin out of four wooden posts with some chicken wire wrapped around them, or make use of a large bin bag with some holes punctured in the sides to allow fresh air to circulate. These are easy ways of making use of autumnal leaves that so many people find problematic.

6.2 Worm composting

Worm composting systems, 'wormeries' or 'vermi-composting', have been widely marketed to compost food residuals in containers, and are particularly useful for households who don't have direct access to a garden. Unlike leaf mold composting, wormeries rely almost exclusively on Greens.

The worms are those that prefer to live in decaying organic matter, and not the earthworms we typically see in our gardens that live in the soil. They can be bought from angling shops, where they are variously called 'brandling', 'tiger' or 'red' worms, although a few other species can also be used.

Like traditional composting, microbes still play a pivotal role in worm composting, as they are primarily responsible for decomposing the Greens. These microbes then, in turn, become the source of food for the worms to eat. As the worms burrow through the decomposing foods, they help create air channels, stopping the wormeries from smelling. The worm casts (poop) are nutrient-rich and biologically active, and make an excellent bio-fertilizer.

The key to effective worm composting is to ensure that the moisture levels are sufficient – not too dry so that the worms dry out, and not too wet so that the air channels fill with water and the composting materials start to smell. Similarly, temperature is also important – not too cold so that the worms become sluggish, and not too warm that they start to 'cook'; room temperature is perfect.

There are a number of different worm composting bins available to buy. Some have a tap so that the liquid fertilizer can be removed easily. It's always best to follow the instructions given by the manufacturer, as these will have been tried and tested to suit the individual system.

6.3 Bokashi composting

Bokashi composting is also suited to food wastes, and, like worm composting, it can be carried out indoors in specially designed buckets. However, rather than relying upon indigenous microbes that colonize the waste materials from the surrounding soil or air, this composting method uses proprietorial brands of microbes: so-called 'Effective Microorganisms'. These microbes have been specially blended to 'ferment' the food waste, although the underlying science is still poorly understood.

Again, if this is of interest a number of proprietary kits are available to buy.

6.4 High-fiber composting

This method was developed during the 1990s by the environmental charity, the Centre for Alternative Technology (CAT) in Wales, UK. Looking into the reasons why people struggled to manage their compost heaps effectively, the team at CAT reviewed which composting materials people had access to. They developed a blend of materials based on food waste and scrunched up cardboard or paper, with the food waste providing the Greens, and the cardboard / paper providing the Browns. Shredding or scrunching up the paper / card also ensures that there is sufficient structure within the heap for air to circulate.

This method is suitable for people with small gardens who wish to compost their kitchen food waste, but do not have access to a steady supply of Browns from their garden throughout the year.

6.5 Food digesters

Some compost bins have been designed specifically to accept food wastes only. Most are aimed at commercial and industrial users; however, some are marketed at householders. An example of a garden digester is the 'Green Cone' which looks like a normal compost bin, but has a basket that sits in the soil (rather than on top of it).

The aim of this unit is to allow soil microbes, worms and insects access to the food, where it is broken down and absorbed into the soil. They do not aim to produce compost, but simply to dispose of food waste.

6.6 Summary

This chapter has summarized some alternative composting methods that differ from the mainstream. Although they all follow the same basic composting principles described in this book, some focus on different kinds of composting materials, such as leaves or high-fiber feedstocks, or rely upon worms or proprietary microbes.

7 Putting Your Compost to Work

Now you've mastered the art of making compost, you'll naturally want to start using it. But how does it differ from the compost you buy in the shop and how can you use it? To try to answer these questions, I've attempted to describe some of the basic properties of compost and the uses garden compost can be best put to. This is a fascinating subject and one that deserves much more attention than a single chapter in a book; however, I hope that it will provide you with sufficient information to empower you to put your compost to work.

7.1 The benefits of using compost

Organic matter

The main benefit of compost comes from the fact that it is rich in organic matter. In Chapter 3 we looked at how the carbon in waste materials is broken down to release both energy for the composting microbes, but also to create a rich, brown substance called 'humus'. (Not to be confused with hummus, the chickpea-based food!). This is an important part of the carbon cycle, which is shown in the box below.

Humus is an important part of soil. Importantly, it acts like a sponge, helping to soak up water and stop plant nutrients from being washed away. Humus-rich soils therefore don't dry out too quickly, as the organic matter helps conserve moisture during periods of drought, as well as reducing soil erosion during times of heavy rainfall.

The organic matter also helps act as a 'scaffolding' within the soil, supporting the sand, silt and clay particles that form the basis of most soils on earth. It helps create 'pores' in the soil, which provides structure to the soil. This helps reduce compaction, whilst also creating small channels for roots to grow into and worms and other soil creatures to tunnel through.

Finally, the organic matter acts as a food source for soil microbes and invertebrates, helping to maintain a living soil that is biologically diverse. This helps reduce the incidence of soil-borne diseases, and also encourages the slow release of nutrients within the soil for plants to use.

Composts made out of predominantly Browns will tend to have a lot of humus in it, and will provide lots of organic matter. On the other hand, composts made out of predominantly Greens (such as grass clippings and food residuals) will tend to have less humus, but will generally have higher nutrient contents.

THE CARBON CYCLE

The building blocks of all life on earth are based on the element carbon, or 'C' for short. This carbon is extracted from the air as carbon dioxide (CO_2) and is 'fixed' by plants into complex carbon molecules using the energy from sunlight through a process called **photosynthesis**. Photosynthesis creates plants, including their stems, trunks, leaves and flowers.

In many respects, composting is the opposite of photosynthesis. During composting, microbes decompose the complex carbon compounds in the plants, using some of it as a source of energy, whilst converting the rest into compost. The carbon that is used as an energy source is released back into the atmosphere as CO_2, whilst the carbon in the compost is converted into 'humus' through a process called 'humification'.

Put simply, composting effectively speeds up what would happen in nature – dead plant matter is partially converted back into CO_2 as it decays, leaving a material that is beneficial to the soil and any plants living in it.

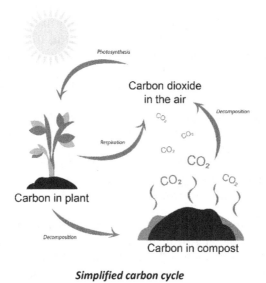

Simplified carbon cycle

Nutrients

Composts contain valuable plant nutrients, both macro-nutrients (that is, those nutrients they need in relatively large amounts, such as nitrogen, phosphorus and potassium) and micro-nutrients (which are nutrients they only need in very small amounts, such as copper and zinc). Depending upon what the compost was made out of, these nutrients will be available at different levels and in different proportions. As a rule of thumb, compost made out of predominantly Brown woody materials will tend to have lower nutrient levels than compost made out of predominantly Greens. This is worth bearing in mind if you have any specific applications for your compost in mind.

One of the main benefits of applying compost to soils is that some of the nutrients are soluble and readily available for plants to use, whilst some are bound up with the humus and are released slowly over a number of years. This means that by applying compost, you will be creating a 'nutrient bank' within your soil, providing a slow-release food for both your plants and soil organisms.

Typical levels and availability of the three main plant macro-nutrients in compost are summarized in the box below.

MAIN PLANT NUTRIENTS IN COMPOST	
NITROGEN	**Effect on plants:** Needed to promote the growth of leaves and stems.
	Typical levels: Between 0.5 – 1.5% (by dry weight).
	Availability: Up to 20% in the first year, then about 8% in every subsequent year.
PHOSPHORUS	**Effect on plants:** Needed to promote the growth of roots and shoots.
	Typical levels: Between 0.3 – 0.4% (by dry weight).
	Availability: About 50% will be available in the first year, with the remainder released thereafter.
POTASSIUM	**Effect on plants:** Needed to promote the growth of flowers and fruit.
	Typical levels: Between 0.5 – 0.8% (by dry weight).
	Availability: About 80% is available to plants in the first year.

Liming effect

Most composts (unless they are made out of certain feedstocks, such as bracken or pine needles) will tend to be slightly alkaline. This means that they have a small liming effect, helping to counteract soil acidity. As some plant nutrients can be 'locked up' in acid soils, applying compost can help release these nutrients.

Good microbes

As compost is made through the actions of countless billions of bacteria, fungi and invertebrates, it therefore contains a rich assortment of living organisms. When added to topsoil, these organisms continue their important work of releasing nutrients and maintaining healthy soil ecosystems.

There is good scientific evidence to suggest that compost can help act as a natural pesticide, helping to suppress certain disease causing organisms that can harm plants, in particular root crops. This is because the beneficial microbes tend to out-compete the disease-causing microbes. The actual means by which they achieve it are complex, and the subject of some fascinating scientific research.

7.2 Bag bought compost

Before we go on to explore some of the ways in which compost can be used, it's probably a good idea to take a look at the products sold in bags through garden centers, home improvement (DIY) stores and the like. They differ from garden compost in some fundamental ways, so understanding this will help prevent you from feeling disappointed at your home-made endeavors.

Bagged compost products are primarily sold as 'growing media' (also called a 'potting soil' or 'potting compost'), which means that they have been developed for use in containers, such as plant pots. They have been specifically blended to provide healthy roots and leaves for plants grown in restricted spaces.

The first major point to note is that not all of these 'composts' sold over the counter contain compost, that is material that has been through a composting process. Until relatively recently, the majority of these products were based on peat dug up from peat bogs.

However, as this causes significant environmental damage, the use of peat in horticultural growing media is being gradually phased out and replaced by alternatives, such as processed bark, steam-treated wood, coir (coconut shell fibers) or composted green waste from large-scale composting facilities. Irrespective of the source, these products have been screened to very small sizes and carefully blended to provide plants with a medium that holds sufficient moisture, whilst allowing the roots sufficient air to breathe.

Secondly, these media have been blended with nutrients (some readily available and some slow release) so that plants continue to grow and thrive over many weeks and months. The specific type of fertilizer will depend upon the product.

Thirdly, the pH (that is, the acidity or alkalinity) of the medium may well have been adjusted, so that it is in the optimal range for most plants. Ericaceous composts, in particular will be slightly acidic so that they favor lime-hating plants.

Finally, some growing media may also contain substances that help it retain moisture and prevent the medium from drying out too quickly.

Please bear in mind that these 'composts' have been subjected to many years of intense scientific research so that they contain the most appropriate blends for their intended purpose. Homemade garden compost differs substantially, so it is unrealistic to expect it to perform in the same way as commercial products.

7.3 Compost applications

Now that we have taken a brief look at some of the benefits of using compost and of some of the properties of bag bought products, we need to now look at some of the ways in which the garden compost you produce can be put to best use. In essence, there are two main ways of using compost: either as a **soil improver** or a **growing medium** (potting soil or potting compost). Both are described below, along with some suggested application rates.

Soil improvers

Soil improvers are materials that are added to soil with the primary aim of improving the soil's physical properties, although they nearly always improve soil chemistry and biology as well. Using compost as a soil improver is largely dependent upon its organic matter content, which, as we've discovered, helps prop up the soil's network of pores that channels the air and moisture needed for healthy plant growth.

Soil improvers can be used in a number of different ways, such as digging it into the soil, or layering it on the surface. Some of these applications include:

- **Soil conditioners** – where fine compost is dug into the top layers of soil to help improve the overall properties of the soil. It is particularly useful in heavy clay soils, which tend to easily compact, and light sandy soils, which dry out quickly.
- **Surface mulches** – where coarse compost is spread on the surface of the soil to help retain water, reduce the growth of weeds, and help protect plants during cold weather.
- **Planting media** – where compost is placed in the bottom of a pit into which trees, shrubs and other plants are placed. This helps the plant establish a new root system and provides it with some nutrients to use in the meantime.
- **Turf dressing** – finely sieved compost can be spread on top of lawns to provide nutrients, help reduce compaction and moss growth, and promote more vigorous grass.
- **Topsoil** – fine compost can be blended with soils that are low in organic matter to create 'topsoils' that can be used for a variety of purposes.

How much compost you need to use for each of these different applications, will, of course, depend upon how much you have available, the type of soil you have in your garden, the types of plants you are growing, and how much rainfall and sunlight your garden receives. Some suggested application rates are shown in the box below. Please bear in mind, however, that these are simply suggestions based on typical applications – using a bit more compost or a little less may also work equally as well, and may even be necessary under certain circumstances.

Using compost as a soil improver

SOIL CONDITIONER

Borders with established plants

Sprinkle compost about 0.5 - 1 in (1 - 2.5 cm) deep onto the soil, then dig this into the top 4 - 8 in (10 – 20 cm) of soil, so that the compost is incorporated into the root zone. This is approximately equivalent to a wheelbarrow load for every 6 - 12 yd^2 (5 - 10 m^2).

Vegetable plots

The nutritional needs of vegetables and soft fruit should to be taken into account, as should the type of soil you have. Some plants, such as potatoes, need well fertilized soils to thrive, whilst others are less nutritionally demanding. It's always best to apply compost just before planting out seedlings or sowing seeds, so that the soluble nutrients don't get washed away by the rain.

Depending upon your soil, you may need to apply anywhere up to 4 in (10 cm) deep of compost, digging it into the top 4 - 8 in (10 - 20 cm) of soil. Well fertilized, organic matter rich soils may only need an inch to replace nutrients used up by the previous crop. Poor soils may well need extra organic matter – if you need the organic matter but don't want to over-fertilize, consider using well-rotted leaf mold, as this is perfect for the job.

Poor soils (especially heavy clay and light, sandy soils)

Some soils are particularly low in organic matter, meaning that they may either compact easily (such as clay-rich soils) or dry out very quickly (such as sandy soils). Either way, a good dose of organic matter-rich compost can begin to ameliorate some of these problems. How much to add? Well, anywhere up to 1 part compost mixed into 1 part soil (by volume). Coarse compost is particularly good, as the twigs and other woody material will provide good structure and will also decompose slowly in the soil over time, continuing to add much-needed humus.

PLANTING MEDIUM

You can give potted plants, especially trees, a helping hand by using compost when planting outside into the soil. Dig a pit large enough to accommodate the roots, then add a shovel load of compost into the bottom of the pit before inserting the plant and backfilling. The aim here is to help the plant recover and establish a new root system. Take care not to provide it with too many nutrients, otherwise the plant will have little reason to set down a strong network of roots. The volume of compost should be approximately one quarter to one third of the volume of soil that will be backfilled around the tree.

SURFACE MULCH

Weed suppression and water retention

This is where twiggy, coarse compost comes into its own. Mulches work by creating a physical barrier over the soil, that then either prevents light from entering (and hence weeds growing), or reduces the rate at which moisture evaporates (thus preventing the soil from drying out too quickly). Generally, a layer of coarse woody compost spread anywhere between 4 and 8 in (5 and 10 cm) thick should work well. Take care to keep the mulch an inch or so away from the stems of plants.

Under fruit trees

Mulching underneath fruit trees and shrubs is an ideal way of providing your crops with a slow release fertilizer, as well as helping retain moisture during the summer months. Mulching between 4 and 6 in (10 and 15 cm) of compost for about 0.5 yd² (0.5 m) around the plant has been shown to work well in commercial crops.

TURF DRESSING

Sprinkle up to 0.5 in (1 cm) deep of finely sieved compost onto lawns, then either brush or rake it in so that it is evenly distributed. Heavy or waterlogged lawns can also benefit from compost if it is mixed with sharp sand (the sort builders use). This can help improve drainage and reduce the growth of moss.

TOPSOIL

Blending poor quality sub-soils with compost can create artificial topsoils. Large-scale trials on degenerated land suggest that a mix of between a quarter and a third of compost mixed with soils low in organic matter works best.

Growing media (potting soil or potting compost)

A growing medium is a material, other than soil, in which plants are grown. It is also sometimes called a 'potting soil' or 'potting compost'. In common parlance, it generally refers to bag bought 'multi-purpose compost', which is used to raise and grow plants in a variety of containers such as seed trays and plant pots. Homemade garden compost doesn't usually have the required physical and chemical properties to be used successfully as a growing medium, especially on its own. Instead, it needs to be sieved carefully, then mixed with other ingredients.

The key characteristics of any growing medium are shown below:

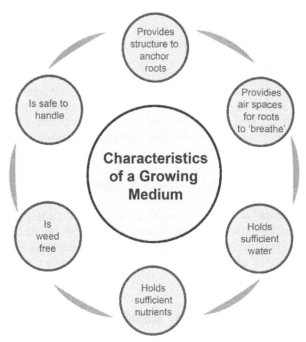

Characteristics of a growing medium

The specific characteristics of any compost will, of course, depend upon the materials it is made out of. So, if your compost has been made out of a lot of Greens, chances are that it will have a much higher nutrient content than a compost made mostly out of Browns. This intrinsic variability is what makes using garden compost in containers a bit hit-and-miss. (Remember that bag bought products have been extensively formulated and tested to deliver just the right level of nutrients, over a specific period of time.)

To overcome this, the best compost to use in a potting mix is one that has been made out of leaf mold. This will have a much lower nutrient content than general garden compost, is less likely to contain weed seeds, and, if it has been left for long enough, should have a fine, crumbly texture. If you plan to use garden compost then only consider using well-rotted, mature compost, as fresh, or actively composting material may well harm plants, something that will clearly be counter-productive. It will also need to be well sieved, so that all large fragments are removed; this is a job that can be quite time consuming.

Mixing your own growing medium can therefore be a leap into the unknown. If you still wish to have a go, then the best thing to do is experiment, bearing in mind that the main factors to consider are the physical structure of the medium (how well it holds water and doesn't compact), as well as its nutrient content and availability. Generally, media used to raise seeds and cuttings should not be too nutrient rich, as high levels can damage young plants.

Mixing leaf mold and garden compost together is an option, although you may need to consider adding some other amendment to provide additional structure, such as sharp sand, perlite or vermiculite. Some suggested mixes are shown below, based on volumes of materials (not their weight).

Using compost as a growing medium

GENERAL POTTING MIX
▦ 1 part sieved compost;
▦ 1 part soil; and
▦ 1 part sharp sand, perlite or vermiculite.
OR
▦ 1 part sieved compost;
▦ 1 part sieved leaf mold; and
▦ 1 part soil.

PROPAGATING SEEDLINGS AND CUTTINGS

▓ 1 part leaf mold; and

▓ 1 part sharp sand, perlite or vermiculite

Commercial growers and dedicated gardeners will only use soil that has been heated (pasteurized) to kill off any weed seeds or disease-causing organisms. This is a fiddly process that uses up energy, so only attempt if you are really, really keen! Place the soil on a baking tray and cover with baking foil, then heat it up to 180 °F (80 °C) for 30 minutes, or 160 °F (70 °C) for 1 hour.

THE DIFFERENCE BETWEEN PERLITE AND VERMICULITE

Both perlite (a type of volcanic glass) and vermiculite (a type of silicate) are non-organic materials that can be added to a growing medium to adjust its physical properties. Both are pH neutral and don't have any nutritional content. They both help retain water within the mix and provide aeration, but they differ in two important ways:

▓ Vermiculite is better at retaining water than perlite; whilst

▓ Perlite provides better aeration than vermiculite.

The type of compost you have and the type of plants you wish to grow will dictate how much perlite or vermiculite you need to use.

Please note that if you plan to grow lime-hating (ericaceous) plants in your potting mix, then do this with caution. As most compost is naturally alkaline, this could potentially damage your plants. Compost made out of bracken or pine needles tends to be slightly acidic, so these could make useful growing media for ericaceous plants.

Specialist compost applications

Compost teas

A compost tea, or steepage, is made by suspending compost in a permeable bag in water, then leaving it to soak for anywhere from a few hours to a few days. Depending upon the recipe, regular stirring may or may not be required, as some teas require oxygen and others don't.

The name 'compost tea' suggests that this is something you may concoct to give to your plants to drink as a fertilizer. Whilst this is one potential application, a more interesting one is to use the tea as a spray to help control leaf-borne diseases. A lot of the research looking into the effectiveness of compost teas has been carried out using composted animal manures, which tend to be richer in nutrients than ordinary garden compost. It goes without saying that if you plan to make a compost tea to use as a fertilizer, then you'll need to use a compost rich in nutrients. If you plan to make a tea to spray onto leaves, then having one rich in composting microbes is important, as it is thought that these microbes outcompete those that cause disease, helping the plant to recover from infection.

If you wish to make a compost tea, then here is a recipe you may wish to try:

RECIPE FOR SIMPLE COMPOST TEA
Mix:
▪ 1 part (by volume) of well-rotted, mature compost with
▪ 1 part water.
Mix well and leave it to steep for 24 hours, mixing occasionally.
Drain the liquid into a watering can, leaving as much of the compost solids in the bucket as possible, then use as a liquid feed or foliar spray.

Please also remember to follow the advice on safeguarding your health in the next chapter, and try not to breathe the spray droplets, as they can have a high microbial load.

Worm cast compost

Worm casts are really just worm feces (poop) and they make up the majority of worm compost. Worm compost, therefore, tends to be high in nutrients and have a fine, crumbly structure. As the worms that produce it have been fed on food waste, it is probably low in weed seeds and other plant disease-causing organisms. It can also be harvested as both the solid casts (compost) or as a liquid that has drained to the bottom of the wormery; some wormeries have a tap at the bottom specifically to enable this.

It is best to consider worm compost primarily as a fertilizer, rather than a soil improver. Because of this, it would be unwise to use worm compost neat, but consider using it as a top dressing or a component within a growing medium mix.

Using worm casts

SOLID WORM CASTS
▓ Apply it to the surface of the growing medium of potted plants to provide additional nutrients; or
▓ Use it as part of a potting mix for nutrient-hungry plants, such as tomatoes.

LIQUID
Dilute 1 part of worm liquor into 10 parts water and use as a liquid fertilizer.

7.4 Summary

In this chapter we've looked at some of the benefits of using compost. Compost helps add structure to soils, improving its resilience to drought and flooding; it creates a nutrient bank for plants to tap into overtime; and it also helps increase biological diversity in soils, as well as reducing acidification.

Homemade garden compost can be put to a variety of uses, although it is best applied directly to soil as a soil improver, either dug into the soil, or layered on the top as a surface mulch. Specialist mixes of potting compost can also be made out of garden compost; however, these are more technically demanding and may require a bit of trial and error. Leaf mold compost is particularly suited to blending into a growing medium; just don't expect it to behave in the same way as commercially-produced media. Finally, worm cast compost and compost teas are some of the more fascinating ways in which compost can be put to work.

8   Health and Safety

It would be remiss of me to write a book on composting without making reference to you, the composter. Although making garden compost is a fairly low risk activity, it's sensible to take some simple precautions to stay safe and healthy. I've included some advice here in this chapter to help point you in the right direction. Most of it is common sense, with some of the suggestions adapted from large-scale commercial processes where health and safety laws apply.

8.1   Protecting your hands

Composting feedstocks often have thorns or other sharp pieces that can puncture or graze skin. The longer these have been left to compost, the greater the chance they may cause infection, simply because there will be more microbes attached to them. It's best to prevent this happening in the first place by taking these few simple measures:

- **Wear gloves when handling feedstocks or compost** – puncture-resistant gloves are particularly useful when handling thorns or other sharp material. Not only do they help prevent cuts and abrasions occurring in the first place, but they also place a barrier between your skin and any irritants that may be present in the sap or soil. Some plants, such as chrysanthemums, ivy and Euphorbia are known to irritate skin, so prevention is always better than cure.
- **Cover up any cuts or abrasions with a waterproof Band-Aid / plaster** – this helps prevent any microbes from entering your skin where they could cause an infection.

If you are pruning, lopping, chopping, chipping or shredding, the chances are you will be using any number of implements aimed at cutting through plant material (such as secateurs, loppers and saws). The key here is to ensure you only cut through the plant material and not your fingers or hands. It sounds obvious, but I'm speaking from experience here! Wearing stout gloves can reduce the risk of serious harm to your fingers and hands. If you are using a motor-driven chipper or shredder, then always follow the manufacturer's instructions carefully.

8.2   Protecting your back

Composting inevitably involves lifting – whether that be feedstocks into the bin, or compost out of it – so a certain amount of manual labor is inevitable. Taking simple precautions can help prevent back pain and damage to your spine. These include:

- Starting with a good posture and directly facing the material you plan to move;
- Keeping your back straight whilst lifting; and
- Trying not to twist around when carrying a heavy load.

8.3   Protecting your eyes and face

If you are chipping or shredding, then there is a risk that small fragments of plant material may be ejected at high speed from the machine, which could cause eye or skin damage. During these activities it's best to wear a full face visor or safety glasses. Again, the manufacturer's instructions should always be followed.

8.4   Protecting your health

Preventing infections

In healthy people, infections in the garden generally occur in two main ways: through the skin, or by swallowing something infectious. We have already discussed protecting your hands by wearing gloves and covering up injuries to prevent infections from entering the skin. However, if you are handling composting materials there is always the possibility that infectious germs (called 'pathogens') can remain on your hands and transfer themselves to your food and drink, only to enter your mouth and gut. It is important, therefore, to remember to follow good hygienic practices by:

- Always washing your hands thoroughly with soap and water after making or using compost, especially before eating or drinking.

This is particularly important if you have been moving manures, as they are teeming with potentially harmful bacteria. If you don't have access to running water near your composting patch, then consider using an anti-bacterial hand gel. (Remember also never to put cat or dog feces in your compost heap, as these can contain parasites that can infect people.)

If you have composted manures, or if you suspect that the compost heap may have suffered a rat infestation at some stage, then it is wise to only use the compost either on non-edible plants, or on vegetables that need cooking first. Ready to eat crops, such as salads, may potentially come into direct contact with pathogens, so care needs to be taken.

Tetanus

Thankfully the specter of 'lockjaw' that haunted our ancestors has long since passed. However, the microbe that causes lockjaw, or tetanus as it is now called, remains just as prevalent. Tetanus is caused by a bacterium that lives naturally in the soil, only causing us problems when it enters broken skin, especially through deep cuts. We can protect ourselves by ensuring that our tetanus vaccinations are up to date and that any cuts are cleaned and protected with a waterproof Band-Aid / plaster immediately.

Protecting your breathing

Moving composted materials around can liberate dust and the very microbes that do the composting, causing them to become suspended in the air. These airborne microbes are termed 'bioaerosols' and include the cells of fungi and bacteria. They occur naturally and we all inhale them every day, even in our homes; however, when present at high levels, they can lead to breathing difficulties in some people. People who have existing chest conditions, or who have impaired immune systems are most at risk.

The simplest way to control it is to:

- Dampen down compost with water before moving it around – this reduces its dustiness;
- Only sieve compost outdoors, and not in an enclosed space, such as a shed; and
- Wear a dust mask if you have an existing respiratory problem – but make sure it is clean and fits effectively around your mouth and nose.

Research I carried out many years ago showed that by simply watering a pile of compost before turning it, had the effect of reducing the concentration of airborne bacteria by over ten-fold.

Sometimes a bacterium called *Legionella* can grow in stored compost. Again, it lives naturally in soil but can be problematic, especially in the elderly and those with poor immune systems. Its growth can be reduced by storing compost in a cool place, preferably outdoors, whilst exposure can be reduced by watering compost and using it in well ventilated areas, preferably outdoors.

8.5 Summary

Making good compost also means doing it safely and hygienically. This chapter has touched on some of the health and safety essentials, which are summarized in the box below.

HEALTH & SAFETY ESSENTIALS

- Always wear gloves when handling feedstocks or compost.
- Cover up any cuts or abrasions with a waterproof Band-Aid / plaster.
- Always wash your hands thoroughly after making or using compost.
- Dampen down compost before turning or moving it around to reduce its dustiness.
- Always follow the manufacturer's instructions carefully when using chippers or shredders.
- Make sure your tetanus jab is up to date.

---

9   Want to Learn More?

If reading this book has whetted your appetite for further information, then there are a wide range of additional resources available to buy or download from the internet for free. Some of them are excellent, some good, and some less so. I've listed some of the English language books and websites in this chapter, which I feel provide balanced information should you wish to research the subject further. However, as there is a practical limit to the number of resources I can list (including those in languages other than English), an omission here doesn't necessarily imply it isn't worth reading.

9.1   Home composting books

Some of the best books include:

Compost
Ken Thompson (2011) Dorling Kindersley. ISBN 978 1 4053 6229 0

Compostology 1-2-3: Composting Made Simple
Ethne Clarke (2011) Rodale Press. ISBN 978 1 60961 770 7

How to Make and Use Compost The Ultimate Guide
Nicky Scott (2010) Green Books. ISBN 978 1 900322 591

The Garden Organic Book of Compost
Pauline Pears, Heather Jackson, Jane Griffiths and Harriet Kopinska (2011) New Holland Publishers. ISBN 978 1 84773 437 2

The Rodale Book of Composting
Deborah L. Martin (1999) Rodale Press. ISBN 978 0 87857 991 4

9.2 Useful home composting websites

North America

www.compost.org
Information about composting in Canada from the Compost Council of Canada.

www.compostingcouncil.org/home-composters/
This website of the US Composting Council contains a list of useful links to composting resources in different US states.

www.mastercomposter.com
Useful information from the USA.

Australasia

www.compostweek.com.au/core/
Information about Compost Awareness Week in Australia, run by the Centre for Organic Research & Education.

www.createyourowneden.org.nz
Information about composting in New Zealand.

India

www.dailydump.org
Information about composting in India.

UK

www.homecomposting.org.uk
Dedicated home composting website run by the UK's organic gardening charity, Garden Organic.

www.recyclenow.com/reduce/home-composting
Information about home composting from the UK's Waste and Resources Action Programme

www.rhs.org.uk
Information about compost, composting and gardening in general from the UK Royal Horticultural Society.

9.3 Information about compostable plastics

General information

www.en.european-bioplastics.org
Information about European Bioplastics.

www.bpiworld.org
Information about the US-based Biodegradable Products Institute.

Certification schemes

www.bioplastics.org.au/
Information about the Australasian Bioplastics Association and certification.

www.compostingcouncil.org/compostable-logo-project/
Information about the US compostable logo project.

www.organics-recycling.org.uk
Information about packaging and products recoverable through home composting in the UK.

www.okcompost.be/en/home
Information about the OK compost scheme.

9.4 Community and master composting

www.bae.ncsu.edu/topic/composting/pubs/backyard-composting.pdf
Information on how to plan community backyard composting programs.

www.communitycompost.org
Website of the UK Community Composting Network.

http://cwmi.css.cornell.edu/mastercompostermanual.pdf
Useful guide to master composting by the Cornell Waste Management Institute.

www.growingwithcompost.org
This EU-funded project provides information about setting up social economy composting projects across Europe.

www.highfieldscomposting.org/sites/default/files/files/resources/growing-local-fertility.pdf

Link to the helpful guide: Growing Local Fertility: A Guide to Community Composting.

www.homecomposting.org.uk/master-composters-mainmenu-36
Information about master compost programs in the UK.

www.valuingcommunitycomposting.org.uk
This provides a toolkit for unlocking the potential of community composting.

Cooperative Extension Programs
Most state universities in the USA run extension programs to help people improve their lives through research-based knowledge, with many supporting backyard composting programs.

9.5 Compost science

Compost, Science & Utilization
A quarterly peer-reviewed scientific journal that reports on compost production, compost quality and its use.
www.tandfonline.com

BioCycle
Monthly US-based magazine and website on composting, organics recycling, anaerobic digestion and renewable energy.
www.biocycle.net

Cornell Waste Management Institute
This website provides access to a variety of composting educational materials and programs developed at Cornell University. In particular, it contains detailed information about the science and engineering of composting.
http://compost.css.cornell.edu/index.html

Ohio Composting and Manure Management
Information about making and using compost, including copies of course slides.
www.oardc.ohio-state.edu/ocamm/t01_pageview/Home.htm

9.6 Carbon-to-nitrogen ratios

Detailed information about the carbon-to-nitrogen ratios

http://compost.css.cornell.edu/calc/cn_ratio.html
An explanation of what the ratios mean and how they can be calculated.

http://compost.css.cornell.edu/OnFarmHandbook/apa.taba1.html
Table listing the C:N ratio and nitrogen content of common composting feedstocks.

www.sniffer.org.uk
Report on Carbon to Nitrogen ratios in UK compost feedstocks.
(Enter "ER15 " in Project Search field).

9.7 Herbicides in compost

Aminopyralid in manures
Website by an agrochemicals company describing how to avoid aminopyralid in horse manures.
www.manurematters.co.uk

An investigation of clopyralid and aminopyralid in commercial composting systems
Report assessing the risks of contamination through large-scale composting systems.
www.wrap.org.uk/sites/files/wrap/Clopyralid%20Report.pdf

Information from Washington State University
Although the website is now dated, it details problems that arose in Washington State.
http://puyallup.wsu.edu/soilmgmt/Clopyralid.html

Acknowledgements

I am grateful to the following people for their input into the production of this book:

- **Erie Myers** – for his patience and professionalism in transforming my rough, hand drawn images into clear-cut diagrams.
  www.fiverr.com/eriemyers

- **Justin French-Brooks (Word to Dialogue)** – for his tactful editing and helpful suggestions regarding the use of American-British English.
  www.wordtodialogue.co.uk

- **Chas Ambrose** – for his support and advice.

About the author

Dr Jane Gilbert is a chartered environmentalist and waste management professional, who has been involved in the composting sector for over twenty years. She is the former CEO of the UK Composting Association and co-founder of the European Compost Network.

Jane originally trained as a microbiologist, has a doctorate in biochemistry and an MBA. She currently provides consultancy and writing services, and works with national and international non-governmental organizations.

Jane has authored a number of technical composting books and has presented at conferences in North America, Europe, Africa and Asia. She has recently launched Carbon Clarity Press, specializing in publishing resources to inspire sustainable living.

Jane lives in Northamptonshire, England, with her husband, two children and cat. When not running around after them, she likes to spend time in her garden, making compost, of course!

For further information visit: www.carbon-clarity.com

Glossary of terms

Accelerator / activator - A commercial preparation of enzymes and microbes that can kick - start composting.

Aerobic - In the presence of oxygen.

Anaerobic - In the absence of oxygen.

Bokashi composting - Composting method that uses proprietorial brands of effective microorganisms.

Booster - See 'accelerator'.

Browns - Woody, dry feedstock materials that are rich in carbon and low in nitrogen.

°C - Degrees Celsius or Centigrade, a measure of temperature. 0 °C is the freezing point of water, equivalent to 32 °F.

Carbon - A chemical element that is present in all life forms. It has the symbol C.

Carbon - to - nitrogen (C:N) ratio - The relative proportion of the elements carbon and nitrogen. It is an indication of the balance of composting feedstocks.

Compost - A solid particulate material that is the result of 'composting' and that confers beneficial effects when added to soil, used as a component of a growing medium, or used in another way in conjunction with plants.

Compost tea - A steepage made by suspending compost in a permeable bag in water, then leaving it to soak for anywhere from a few hours to a few days.

Compostable - Capable of being composted.

Composting - A natural process in which waste plant and animal materials are broken down to create 'compost'.

Contaminant - An impurity that can taint compost.

cm - Abbreviation for centimeter, a unit of length, equivalent to 0.4 inches.

Decompose / decomposition - The degradation of feedstocks by composting microbes to create compost.

Density - A measure of how closely packed the composting materials are.

DIY - Refers to a 'Do - It - Yourself' store; home improvement store.

°F - Degree Fahrenheit, a measure of temperature. 32 °F is the freezing point of water, equivalent to zero degrees Celsius.

Feedstock - Waste material used in composting.

Fire fang - Spores of some composting microbes; has the appearance of white dust.

Greens - Soft, sappy feedstock materials that are high in nitrogen.

Growing medium - A material, other than soil, in which plants are grown in containers. Also called a potting soil or potting compost.

Hazard - Something that has the potential to cause harm.

Heavy metal – A metal that has the potential to be harmful to the environment.

High - fiber composting - Composting method using food waste and scrunched up cardboard or paper.

Humification - The process of forming humus.

Humus - Decomposed organic material.

in - Abbreviation for inch, a unit of length, equivalent to 2.5 centimeters.

Leaf mold - A low - nutrient compost formed out of autumnal leaves that have been allowed to decompose on their own.

Liming effect - Ability of compost to help counteract soil acidity.

Mesophile / mesophilic - Microbes that prefer to live in mid - range temperatures, somewhere between 68 and 113 °F or 20 and 45 °C.

m - Abbreviation for meter, a unit of length, equivalent to 1.1 yards.

m³ - Cubic meter, a unit of volume, equivalent to 1.3 cubic yards.

Microbe / microorganism - Tiny single - celled organisms that carry out composting, including bacteria and fungi.

Mulch - A coarse compost that is spread on the surface of the soil to help retain water, reduce the growth of weeds, and help protect plants during cold weather.

Nitrogen - A chemical element that is present in all life forms. It has the symbol N.

Odor - An unpleasant smell.

Odorous - Unpleasantly smelly.

Planting medium - A compost that is placed in the bottom of a pit into which trees, shrubs and other plants are placed to assist the plant establish a new root system.

Potting compost - See 'growing medium'.

Potting soil - See 'growing medium'.

Primary consumers - Animals that feed off composting microbes and dead pieces of vegetation.

Risk - The chance or likelihood of being harmed by a hazard.

Secondary consumers - Animal predators that feed off primary consumers.

Soil - A mixture of minerals, organic matter, gases, liquids and organisms on the surface of the ground that together support plant life.

Soil conditioner - A compost that is dug into the top layers of soil to help improve the overall properties of the soil.

Soil improver - Materials that are added to soil with the primary aim of improving the soil's physical properties.

Thermophile / thermophilic - Microbes that prefer to live at high temperatures, above 104 °F or 40 °C.

Topsoil - Top layer of soil, high in organic matter.

Turf dressing - A finely sieved compost spread on top of lawns to provide nutrients, help reduce compaction and moss growth, and promote more vigorous grass.

Turning - The process of picking up partially composted material and then dropping it back down, with the aim of mixing it as much as possible to introduce fresh air and re - create air channels.

Vermicompost - Compost made in a wormery using 'brandling', 'tiger' or 'red' worms.

Vermin - Pests or nuisance animals.

Wormery - Specialized container for composting with worms. See 'vermicompost'.

yd - Yard, a unit of length, equivalent to 0.9 meters.

yd³ - Cubic yard, a unit of volume, equivalent to 0.8 cubic meters.

Index

Made in the USA
Monee, IL
27 March 2020